COGNITION, AGENCY AND RATIONALITY

PHILOSOPHICAL STUDIES SERIES

VOLUME 79

Founded by Wilfrid S. Sellars and Keith Lehrer

The titles published in this series are listed at the end of this volume.

COGNITION, AGENCY AND RATIONALITY

Proceedings of the Fifth International Colloquium on Cognitive Science

Edited by

KEPA KORTA

Institute for Logic, Cognition, Language, and Information (ILCLI), Donostia – San Sebastián, Spain

ERNEST SOSA

Department of Philosophy, Brown University, Providence and Rutgers University, New Brunswick, U.S.A.

and

XABIER ARRAZOLA

Institute for Logic, Cognition, Language, and Information (ILCLI), Donostia – San Sebastián, Spain

KLUWER ACADEMIC PUBLISHERS

DORDRECHT / BOSTON / LONDON

A C.I.P. Catalogue record for this book is available from the Library of Congress.

ISBN 0-7923-5973-9

Published by Kluwer Academic Publishers,
P.O. Box 17, 3300 AA Dordrecht, The Netherlands.

Sold and distributed in North, Central and South America
by Kluwer Academic Publishers,
101 Philip Drive, Norwell, MA 02061, U.S.A.

In all other countries, sold and distributed
by Kluwer Academic Publishers,
P.O. Box 322, 3300 AH Dordrecht, The Netherlands.

Printed on acid-free paper

Printed in the Netherlands.

TABLE OF CONTENTS

INTRODUCTION vii

ARE HUMANS RATIONAL?
Ernest Sosa 1

HOW CAN FALLACIES ARISE ABOUT FALLACIES?
L. Jonathan Cohen 9

RETHINKING RATIONALITY: FROM BLEAK IMPLICATIONS TO DARWINIAN MODULES
Richard Samuels, Stephen Stich, and Patrice D. Tremoulet 21

THE PSYCHOLOGY OF COUNTERFACTUAL CONDITIONALS
Ruth M.J. Byrne, Alessandra Tasso, and Valerie Thompson 63

MEANINGS AND CONCEPTS
Stephen Schiffer 79

CONTEXTUALISM AND THE MEANING-INTENTION PROBLEM
Thomas Hofweber 93

PRESUPPOSITIONAL AND RHETORICAL EFFECTS OF FOCUS PARTICLES: THE CASE OF 'EVEN'
Fernando García Murga 105

LOGICS FOR MENTAL STATES
Ton Sales 123

SOCIALLY RESPONSIBLE DECISION MAKING BY AUTONOMOUS AGENTS
Susanne Kalenka and Nicholas R. Jennings 135

DIVERSITY IN RATIONALITY. A MULTI-AGENT PERSPECTIVE
Rosaria Conte 151

THE CONTEXTS OF COLLABORATION
Barbara J. Grosz 175

SUBJECT INDEX 189

NAME INDEX 195

INTRODUCTION

COGNITION, AGENCY, AND RATIONALITY

The Fifth International Colloquium on Cognitive Science (ICCS-97), co-organized by the Institute for Logic, Cognition, Language, and Information (ILCLI) and the Department of Logic and Philosophy of Science, both from the University of the Basque Country, took place at Donostia - San Sebastián from May 7 to 10, 1997, with the following as its main topics:

1. Rationality and Psychology of Reasoning.

2. Knowledge in Context: Language, Mind, Society.

3. Rational Agency in Multi-Agent Systems.

4. Compositional Semantics and Natural Language.

A little less than a hundred researchers from all over the world exchanged their most recent contributions to Cognitive Science in an exceptionally fruitful and friendly atmosphere. This volume contains a small but representative sample of the main papers. They all were invited papers except the one by Ton Sales, which corresponds to a tutorial, and the one by Fernando García Murga, a contributed paper that merited the IBERDROLA Best Paper Award, given in ICCS-97 for the second time.

In the first provisional program of ICCS-97 a tutorial on "Towards a (psychological) Pragmatics" was scheduled, to be delivered by Víctor Sánchez de Zavala, Emeritus Professor of Philosophy of Language at the Institute and the Department here, in Donostia - San Sebastián. At the end of October 1996, we were much affected by the sudden news of his death. His absence was especially and deeply felt by his colleagues during the Colloquium, which the organizers dedicated to his memory. We want to do the same with this volume.

The volume contains three kinds of papers corresponding to three of the main disciplines in Cognitive Science: philosophy, psychology and artificial intelligence. Sosa's, Cohen's, Schiffer's, and Hofweber's are clearly philosophical papers; Samuels, Stich and Tremoulet's and Byrne, Tasso and Thompson's are within psychology; Sales's, Kalenka and Jennings's, Conte's and Grosz's are from Artificial Intelligence. The title "Cognition, Agency and Rationality" is intended to capture the main issues addressed by the papers. Of course, all are concerned with cognition, but some are specially centered on the very concept of rationality (as Sosa's, Cohen's, Samuels, Stich and Tremoulet's or Conte's) while others are focused on (multiple) agency (specially, Kalenka and Jennings's and Grosz's

K. Korta et al. (eds.), Cognition, Agency and Rationality, vii-xi

papers). García Murga's paper lies somewhere outside these groupings: it is a paper in linguistics (semantics and pragmatics) and addresses the particular topic of presuppositions.

Now, we will briefly describe each paper one by one.

1. Ernest Sosa, 'Are Humans Rational?'. A stream of experimental results has put in doubt the traditional conception of man as the rational animal. The mistakes people make are said to be more than just occasional and superficial. They are said to be systematic, entrenched, and as deep as misapplying modus ponens in propositional logic and the conjunction rule in probability theory. Some have tried to explain away the apparent mistakes as deriving from misinterpreted instructions. On a case by case basis, this sort of interpretational response can seem quite plausible. On a broader view, though, it becomes increasingly unattractive. As each new interpretative proposal runs afoul of the next batch of experimental results, the reinterpretation strategy starts to seem disturbingly ad hoc. But what do these experimental results really show? What exactly is rationality and just how do the results show us to fall short of it?

2. L. Jonathan Cohen also addresses the issue of the consequences of the experimental results on human rationality in his 'How Can Fallacies Arise about Fallacies?'. These experiments that purported to show 'bleak implications for human rationality' came under criticism by the author himself and others in the eighties. They argued that psychologists had involved themselves in fallacious reasonings that made them conclude, incorrectly, that the subjects of the experiments were involves in patterns of fallacious reasoning. "How can it be that a considerable number of professionally trained experimental psychologists have all committed similar fallacies in interpreting their subjects' responses, and what is the nature of their mistake?". This is the question that Cohen tries to answer. If the answer is satisfactory, "the case against the 'bleak implications' psychologists is thereby strengthened." He considers the experimental data concerning both judgements of probability and judgements of deducibility, and claims that psychologists are guilty of the fallacy of undisambiguated probability and undisambiguated deducibility. He argues for a pluralistic stance in the conceptual analysis of probabilistic and deductive reasoning.

3. "Rethinking Rationality: From Bleak Implications to Darwinian Modules" by Richard Samuels, Stephen Stich and Patrice D. Tremoulet is the longest paper in the volume. It too addresses the "bleak implications" issue, but in a way more extensive and sympathetic to the empirical side of the debate. After sketching some of the better known experiments in the area, they focus on the notion of competence, which is usually invoked by the "bleak implications" party. One of the most recent challenges to this pessimistic view on human rationality comes from the interdisciplinary field of evolutionary psychology, which defends a highly modular view of human mind. The two main sections of the paper are devoted to the modular picture of the mind advanced by evolutionary psychologists, and to several recent studies that appear to confirm its main predictions, respectively. They conclude that the arguments and the experimental evidence offered do not constitute a conclusive case for the evolutionary psychologists' theory about the mind, and they go on to consider a hypothetical question: "If the evolutionary psychologists'

account turns out to be in the right track, what implications would this have for questions about the nature and the extent of human rationality or irrationality?"

4. How do people reason about what might have been? Ruth Byrne, Alessandra Tasso and Valerie Thompson answer this question in their "The Psychology of Counterfactual Reasoning." First, they present the challenge posed to a general theory of conditionals by counterfactual conditionals, and then they propose a mental model theory of conditionals that covers both reasoning about matters of fact (factual conditionals) and reasoning about matters of possibility and impossibility (counterfactual conditionals). Their theory explains why counterfactual conditionals seem to mean something very different from factual conditionals, even if there is a single cognitive mechanism involved. They also describe some experiments that corroborate the predictions of their theory about the inferences people make.

The next two papers lead us to central topics in the Philosophy of Language and the Philosophy of Mind.

5. "Meaning and Concepts" by Stephen Schiffer extends his case against compositional semantics, by developing a theory of meaning called *pleonastic Fregeanism,* in which the propositions we believe and assert are pleonastic propositions composed of pleonastic concepts or modes of presentation. But, while within full-blown Fregeanism the fact that two propositions are different is explained by the fact that they have different constituents, according to pleonastic Fregeanism, two propositions having different constituents is entirely derivative on the conceptually prior fact that the propositions are determined as different. Schiffer concludes that, though languages may have pleonastic compositional meaning theories, they will not explain any of the things theorists have thought they needed compositional semantics to explain.

6. Thomas Hofweber, in his "Contextualism and the Meaning-Intention Problem" addresses the problem of the context sensitivity of knowledge ascriptions. He defends contextualism against Schiffer argument that the contextualist explanation is in contradiction with the Gricean idea that meaning has to be backed by speaker's communicative intentions (*the meaning-intention problem*). Hofweber argues, contra Schiffer, that in knowledge ascriptions there is an aspect of the content of the knowledge ascribing utterance that the speaker is unaware of, that there is what he calls *hidden relativity*.

7. Fernando García Murga's "Presuppositional and Rhetorical Effects of Focus Particles" presents a general account of the semantics and pragmatics of the word 'even', within a new general theory of linguistic presuppositions which, according to the author, 'redeems' the term 'presupposition' from the wastepaper basket where it was thrown in the late seventies. The theory is built using Jackendoff's Conceptual Semantics and treats 'even' as a focal presuppositional construction whose behaviour is similar to other focal presuppositional elements such as contrastive stress. One important result of the paper is the distinction between linguistic and argumentative presuppositions, which cuts across the distinction between semantics and pragmatics of natural language. This paper won the IBERDROLA Best Paper Award.

8. Ton Sales' "Logics for Mental States" is a schematic survey of formalizations of a wide variety of mental states for computational purposes. The content of the paper was intended as a tutorial on the subject, the most relevant literature on the topic is described in a concise and clear way. This paper leads to the last three chapters which concern multi-agent systems in Artificial Intelligence.

9. One of the main problems in multi-agent systems is to reach a satisfactory balance between the exploitation of the conceptual power of autonomous individual agents and the coherent performance of the overall system. Susanne Kalenka and Nicholas R. Jennings' "Socially Responsible Decision Making by Autonomous Agents" presents a framework for characterising social decision and proposes a socially responsible decision making principle (the *principle of social rationality*), which enables agents to exploit interactions with others for their own gain, but which considers also that they can be sometimes willing to do things for the greater good. Three socially responsible decision making functions are identified and used in a multi-agent system for unloading lorries at a warehouse, making a empirical evaluation of their effectiveness.

10. Rosaria Conte, in her "Diversity in Rationality. A Multi-Agent Perspective", examines the "marriage" between AI (multi-agent systems and distributed artificial intelligence), on the one hand, and economic and strategic rationality (game theory, economic decision theory), on the other. She discusses some of the main limits of the concept of rationality offered by the economic theory and proposes various amendments, which essentially amount to the introduction of diversity in the rationality paradigm. She also presents a computational model (MICROdep) to simulate the formation of partnerships in social markets and test some empirical hypotheses concerning negotiation.

11. Barbara J. Grosz's "The Context of Collaboration" concerns the role of contexts in collaborative activity. After briefly reviewing the roles of contexts in language processing, she presents the SharedPlans model of collaborative activity which encompasses the distinction between "intentions-to" and "intentions-that" as well as the concept "intentional context", and that has been used both as a critical constituent of a computational model of dialogue and as the basis for building computer agents that collaborate with one another or with humans. The intentional contexts and the intention operators are used to distinguish between the actions of two types of business school graduates in a well-known example used by Searle to argue against individualistic accounts of collective activity.

In sum, these *Proceedings of ICCS-97* collect the papers by some very well-known researchers in Cognitive Science, which represent leading edge work on *Cognition, Agency and Rationality.* The diversity of their disciplinary origins and standpoints not only reflects the main topics and the range of different positions presented at ICCS-97 but also serves as proof of the richness, fruitfulness, and diversity of research in Cogntive Science today.

Acknowledgements

ICCS-97 was financially supported by the Secretary of State of Education and Culture (Ministerio de Educación y Cultura) of the Spanish Government, the

Secretary of Education of the Basque Government, the Kutxa Foundation, and Iberdrola Electric Company. We would like to thank the members of the Organizing and Program Committees of the Colloquium for their invaluable work before and during the Colloquium. Concerning the preparation of this volume, an anonymous referee and the responsibles from Kluwer, especially Keith Lehrer, Maja de Keijzer, Rudolf Rijgersberg and Jolanda Voogd must be thanked. And, last but not least, Eli Mendizabal and Edurne Atxa (ILCLI) deserve a special mention here for their help in typing and formatting.

Kepa Korta, Ernest Sosa and Xabier Arrazola

ERNEST SOSA

ARE HUMANS RATIONAL?

1.

A stream of experimental results has put in doubt the traditional conception of man as the rational animal. The mistakes people make are said to be more than just occasional and superficial. They are said to be systematic, and as deep as misapplying modus ponens in propositional logic and the conjunction rule in probability theory.[1]

Some have tried to explain away the apparent mistakes as deriving from misinterpreted instructions. Thus people may be assessing not which proposition is likelier in a mere probabilistic sense, but rather which proposition is more credible or plausible. And if the judgments do concern credibility or plausibility rather than bare statistical probability, then people may be guided by considerations of comprehensive coherence; which would help explain their judgments without violating the conjunction rule or any other rule owed to probability theory. As for the apparently wrong choices concerning which cards must be turned over in the selection task, these may stem from interpreting the conditional involved as subjunctive rather than material, and interpreting "testing" as "obtaining evidence relevant to the truth or falsity of that conditional." That would account for many of the choices people make, while entailing no clash with modus ponens.

Such attempts to explain away the experimental results attain at best limited success, however, since too often there remains a residue apparently resistant to any such treatment. But what do these resistant results really show? What exactly is rationality and just how do the results show us to fall short of it?

2.

People are said to be rational, if and only if, performance errors aside, they follow appropriate rules of reason derived from formal sciences such as logic and probability theory. That is to say, to be rational is to have within one's competence correct rules of reasoning appropriately based on the relevant formal sciences.[2]

To so restrict rationality gives it too narrow a focus, however, since in harboring a certain belief one might fall short of rationality in two main ways, at most one of which fits within so narrow a focus. First, one might believe in a way contrary to how one ought believe. Thus one might disbelieve what one ought to believe, or

K. Korta et al. (eds.), Cognition, Agency and Rationality, 1-8.

believe what one ought to disbelieve. That is one way to fall short. But there is also a second way: namely, to believe where one ought to withhold both belief and disbelief. And it is not at all easy to see how the rules that would counsel withholding in given circumstances might relate to theorems of logic or probability. Someone might arbitrarily believe that the stars are even in number, a clear failure of rationality. What rule deriving from logic or probability rules this out? The rule that if you assess the probability of something at 0.5 then you should not believe it? Among the problems with this, the following deserves special mention: what rules out that our versatile believer just assign his favored proposition a probability to suit, say 0.9? Is there a principle deriving from logic or probability that would prohibit this?

Moreover, one's level of rationality is revealed by one's belief management not only in the armchair but also in the marketplace, the law court, and the laboratory. We still focus here on "theoretical" (belief-management, epistemic) rationality; we still leave aside the practical rationality involved in the acquisition, retention, and pursuit of ends. Even having thus narrowed our focus, we still find that someone who believes something crazy in a way that is either unfounded, or founded just on wishful thinking, or superstition, or the like, would fall short of rationality, in a straightforward and natural use of the term; they would fall short of theoretical, epistemic rationality.

To focus properly on the main issues of human rationality raised by our selected psychological literature, however, we must focus even more narrowly, more narrowly than might be suggested by our most common concepts of rationality. We must first narrow our focus away from the practical rationality of managing one's objectives and their pursuit, and towards the theoretical rationality of belief management. But we must focus more narrowly yet, to exercises of "pure reason." And even here we narrow our focus even further, to the implications for one's rationality that derive from injunctions to believe, and from prohibitions that derive from injunctions to disbelieve (to believe the opposite). So we focus on the irrationality of failing to believe that everything is self-identical, for example, and on the irrationality of believing that something is both material and not material, and the like. But we leave aside the irrationality of believing when one ought to withhold, in cases where this derives from no injunction to disbelieve (not even upon considering the proposition in question). Thus we leave aside the irrationality of believing arbitrarily that the stars are even in number, and the like.

3.

We stick to rationality in belief, then, as opposed to choice or decision, and we stick to *pure* rationality, leaving aside the broader theoretical rationality of proper belief management in general. So our relevant rules would be rules of belief formation or retention whose basis is pure reason. Three sorts of rules are worth distinguishing. First, absolute injunctions: e.g., one is to believe that nothing is self-diverse. Second, absolute prohibitions: e.g., one is never to believe the conjunction of any proposition and its very negation. Finally, there might be conditional rules: e.g., that if such and such conditions hold, then one is to believe a certain proposition (related to the conditions in a certain specified way). However, if the rules of reason are supposed to derive from theorems of the relevant formal sciences, such as logic and probability, if they are supposed to be rules of pure reason, then it is not easy to think of plausible conditional rules. Just consider some sample proposals.

The following have been proposed as conditional rules: [3]

AND-ELIMINATION PRINCIPLE: If you believe the conjunctive statement A **and** B, you should believe both the statement A and the statement B.

MODUS PONENS PRINCIPLE: If you believe A and you believe **if** A **then** B, you should believe B.

But even if you *do* definitely believe the conjunctive statement that the moon is made of cheese and Apocalypse is nigh, still most likely you should *not* believe either one of these conjuncts, much less the two of them. And a similar point applies to the second supposed principle. There are indeed more plausible principles in the vicinity, but these take the form of prohibitions:

AND-ELIMINATION PROHIBITION: One must not believe a conjunctive statement A **and** B, without believing each of A and B.

MODUS PONENS PROHIBITION: One must not believe A and believe **if** A **then** B, without believing B.

Aside from a few positive absolute rules, it is easiest to think of normative principles of reason in the form of absolute prohibitions such as these two. It is here that we find a trove of irrationality results. People apparently violate such prohibitions systematically.

Consider first the positive absolute rules. What is their scope? After all, no-one can avoid neglecting infinitely many theorems of logic and probability, which nevertheless does not impugn anyone's rationality. As for the prohibitions, people often enough believe propositions incompatible with theorems of logic, and even this need not reflect on their rationality. Presumably one is not prohibited from believing *any* such proposition, whatever the circumstances, on pain of irrationality. Which theorems and which propositions, then, are those relevant to rationality and what sets them apart? Why is it that rationality requires belief in the law of identity but not belief in Gödel's incompleteness theorem? Why is it that you do not violate rationality if you just make a minor blunder in your truth-table calculation and believe as you should not; or if you trust a well known textbook where it for once goes wrong and gives incorrect information? As a result of this you may end up believing logically incorrect and necessarily false propositions, with no failure of (your) reason.

We must of course put aside mistakes due to misreading, miscounting, or being misled by unreliable testimony. Such mistakes need not count against one's own rationality. A study of human rationality might focus rather on mistakes attributable to one's belief formation or sustainment in the absence of such operations or methods, since rationality pertains rather to the operation of one's own pure thought. Suppose one ponders a question, and is able to answer either directly or after a bit of reasoning not essentially aided by operations or methods like counting, reading, or accepting testimony; or by faculties like perception or memory. If one still makes a mistake when one answers without reliance on any such external aid, then it is after all one's own faculty of reason that is at fault. Alternatively, we might consider also mistakes that do involve external aids without being attributable to them. Such aids might after all be allowed if we make sure that they all function flawlessly, so that any error must be attributed to one's own pure thought itself, and not to the *combination* of one's thought together with any such aid.

In a failure of rationality, one through pure reason (e.g., through pure reflection, or through mere armchair thought) either believes as one ought not or fails to believe as one ought; or else one falls into some error attributable not to external aids, but only to one's reason. Very well, but this still does not tell us what distinguishes the abstract propositions belief in which is required by rationality, from the many other such truths that one is not called on to accept by pure thought. Nor does it distinguish the propositions that we may not rationally accept just on the basis of pure thought, from those which may properly be accepted on such a basis.

4.

What then is involved in our concept of rationality, in the proper operation of pure reason? What sort of concept is this? Is it an ideal concept, perhaps like perfect flatness, which actual surfaces might approach more or less closely without ever reaching? Is it perhaps a concept of an ideal rationality which some approach more closely than others? Great thinkers would approach the ideal more closely, but all thinkers would be spread along some dimension(s) of rationality allowing differences of degree. Perhaps, but I for one find it opaque what might be involved in such perfect rationality. Would it be direct intuitive belief in all formal truths? After all, if one thinker believed thus intuitively and directly (and by nonaccidental means) all formal truths thus believed by the other, *plus some others besides*, he would presumably show greater powers of pure thought and attain thereby a closer approach to perfect rationality. So, again, would nothing suffice for ideal rationality short of immediate belief in all formal truths whatever?

That suggestion puts the emphasis on intuition, however, and leaves aside modes of belief acquisition employed by the rational, even modes involved in pure armchair thought. It is not clear that one's powers of intuitive reason need bear any special proportion to one's powers of *inferential* reason. People might be differentially gifted in the two respects, the intuitive and the inferential. Take someone very low on the intuition scale but very high on the inference scale. What is his place on the scale of rationality? The problem is this: if the ideal is defined exclusively in terms of power of intuition, then inferential reason is denied its due in overall assessments of rationality. But then, again, how are we to define the ideal so as to make due provision for both aspects of pure reason, the intuitive and the inferential?

Compare the Snell vision chart and the associated ratings of people's eyesight. Such ratings are humanity-relative. Roughly, someone with 20/100 eyesight can see only at 20 feet (or less) what those with good eyesight can still see at 100 feet. People with 20/20 vision see at 20 feet what the better among us can see at that distance, but, presumably not farther. (Not the very best of the elite, indeed, since it is possible to have better than 20/20 vision; just those with average eyesight, or the like.) Given this way of conceiving of the normative concepts pertaining to eyesight, it is incoherent to suppose that humanity at large, as it actually is, could have inferior eyesight. There must be some at least among us gifted with noninferior eyesight. Perhaps then humanity could not as such fall short of rationality, any more than humanity could be universally inferior or constitutionally defective in their vision; or universally short, for another example.

But surely Plato's cave is not a conceptual impossibility. Might not humans have been so impaired as to fall short seriously both in respect of vision and in respect of rationality? Surely that is possible, and yet seems conceptually impossible if we *define* low rationality and poor eyesight in the relative ways indicated above. Whether this is really so, however, depends on a subtlety in the definition.

How do we define shortness? Are we willing to say that all Pygmies are very short? Are we willing to say even this: that if the only humans had been Pygmies, then all humans would have been quite short? Are we not in fact willing to say that our medieval predecessors were, with very few exceptions, quite short? How can all this be true if tallness and shortness are relative terms? Here the answer is obvious. Someone can be tall as a Pygmy, or as a medieval, while short relative to contemporary standards.

When we target someone for application of a concept relative in the way of our concept of tallness, and our target occupies a context not our own, that sets up an ambiguity: are we assessing them relative to *our* context, or relative to theirs?

The same holds for rationality. Even if we define rationality as community-relative, the possibility will remain that people generally have fallen seriously short of rationality. Of course they could not all have fallen short relative to their own average attainments, but they might still have fallen short relative to *our actual* average attainments.

In light of these reflections, rationality is plausibly viewed as relative. This is not to say that anything goes, that what is considered more rational is up to the whims of individuals or the conventions of tribes. That is no more so here than it is so with vision or with height. Who sees better and who is taller is not subject to determination by people's individual decisions or community agreements. These are absolute, non-relative dimensions, and just where people fall within these dimensions is an objective matter of fact. Nevertheless, concepts based on such dimensions can usefully categorize people even while being conventionally relative: consider tallness, or *good* eyesight. Among such relative concepts are those that relativize to *our* actual situation, our actual situation in respect of height, or in respect of vision, or in whatever may be the relevant respect. The concept of being rational, or *adequately* rational, is among such relative, normative concepts. Most commonly, when we assess rationality we do so relative to our actual levels of rationality. What does it means to say that someone is rational, then, or *adequately* rational, or at least sufficiently rational not to count as irrational? Perhaps that is only to say that they do not occupy too low a percentile level of *actual*, human rationality. (This assessment need not be in itself already normatively loaded, any more than is the definition of the tall as those *sufficiently* taller than the average. And how else can it be realistic to define tallness? Not just as "taller than the average," of course, since someone can be infinitesimally taller than the average without being tall.)

5.

All that being so, it is conceptually impossible to demonstrate that humans generally are irrational. The most that might be shown is that humans are not as rational as they might be, a less dramatic result. Compare the claim that humans are endowed with lesser eyesight than they might enjoy. They might have had the vision of eagles; perhaps the vision of Superman. So humans are generally deficient relative to eagles, and relative to Superman. But it does not follow that humans generally have *deficient eyesight*, given the ordinary logic of this concept. Someone has deficient eyesight when their eyesight is deficient relative to average human eyesight, as humans actually are. It is thus conceptually impossible that actual humans generally suffer from deficient eyesight (which, again, does not at all preclude that humans *might* have suffered, generally or even universally, from deficient eyesight.)

Are humans generally, in actuality, seriously irrational? That now seems as conceptually confused as it would be to say that humans are generally, in actuality, seriously short. Not that it is confused to think that people *might* have been seriously short, all of them. The confusion resides rather in the thought that for all we know humans are *in fact, in actuality,* seriously short, all of them. Similarly, although it is quite coherent to entertain the possibility that humans might have been generally irrational, what now seems incoherent is the thought that perhaps, for all we know, humans are in fact, in actuality, seriously irrational, all of them.

One possible way to rescue the thesis of actual human irrationality would requires us to reconsider the analogy with vision. According to the medical book I consulted, to have 20/20 vision is to see at 20 feet what people with "perfect" vision see at 20 feet. This suggests a conception of the quality of eyesight as relative to the condition of the organ of eyesight, the eye, together with a conception of an ideally formed and proportioned eye, and perhaps one that is also ideally positioned relative to the other relevant human systems. This would be a conception of an organ with physical attributes, intrinsic and relational, that fit it ideally for the fulfillment of its function. In the case of seeing we can let the discerning of letters on the Snell chart be an adequate test, and so we can more specifically define ideality in the eye as the optimal shaping, sizing, constitution, and positioning of the eye with a view to discerning the letters on the chart as far and as sharply as can be.

That seems questionable, however, as a way of conceiving of good eyesight. For one thing, even if humans *could have had* eagle eyes, this would hardly show that we are all *in fact* suffering from deficient eyesight. We might be tempted here to fall back on a notion of ideality for a *human* eye. So there are certain properties of a perfectly or ideally formed *human* eye, and excellent eyesight is to be defined relative to that. Moreover, we can understand the notion of such an ideal eye, even if actual humans all fall seriously short of it. However, is it really clear what the limits of the specifically human eye could be, or how that might bear on our evaluation of the eyesight of actual humans alive today? Could not the human eye evolve further to the point where it is much more powerful than are human eyes today? And if that were so, could that show that our actual eyesight falls seriously short, that we are all right now "visually deficient"?

The approach suggested, by appeal to the physical and biological properties of the eye, seems questionable at best. The analogous approach to the study of rationality is even less promising. The implication would be that there is somehow an ideal brain that sets a limit to what a human can attain by pure reason. What would play the role of discerning letters at various distances? When one considers the wide range of human ability, with, say, first-graders at one end, and a mathematically intuitive genius, like Ramanujan, at the other, and when one considers the special combinations of aptitudes and ineptitudes that can be found in a genius or in an idiot-savant, there seems little hope of making sense of anything like "the brain with the ideal constitution and positioning." A brain ideal for one set of rational functions might be less so for another such set. Is there a single dimension constitutive of rationality in such a way that one could define rational excellence in terms of how far along that dimension one resides? At present, that has little to be said for it, and seems an unattractive hypothesis.

Even if we could adjust to a multidimensional basis for reason, moreover, it still seems doubtful that our falling well short of the ideal along *each* of the relevant dimensions would show us to be *irrational*. We would fall short in respect of a degree of rationality attainable to someone with our brain capacity, true enough. And this might be of some interest. Compare the following. Perhaps one could operate on human eyes so as to shape them in a way that would improve acuity. This would show that as we are actually constituted we fall short in respect of a degree of acuity attainable by someone with our eyes. And this might be of substantial interest. For example, people might set out to develop and perfect such an operation. But it seems perverse to derive the conclusion that, because of that possible surgical improvement, therefore people are right now "visually deficient," or fall short seriously in respect of visual acuity.

Of course none of this detracts from the scientific interest and importance of the psychological results. It only puts in doubt conclusions drawn by some, especially the dramatic conclusions that cast doubt on actual human rationality in general. The data on how humans actually tend to fall short in respect of rationality are interesting and important in themselves. Their importance does not require that they entail people to be actually, in general irrational. That would be rather like supposing that the results concerning people's susceptibility to systematic visual illusions would not be important unless they supported a thesis of generalized human blindness.

Let us recall, in conclusion, the paradoxes of Greek antiquity, which show already, beyond any reasonable doubt, the limitations of pure thought. In a true aporia we find ourselves drawn by pure thought to assent, and to assent with apparent certainty, to each of a few propositions, which pure thought reveals just as clearly to be logically incompatible. So pure thought shows itself to go astray, and this is not just a casual mistake, as when one adds wrong or calculates a truth value incorrectly. This is deeper and more disturbing, but nevertheless a robust and systematic result. The contemporary results show the debility of human reason to have wider scope than we might have thought, but they do not show it more powerfully or definitively than the ancient paradoxes.

Ernest Sosa
Brown University
U.S.A.

NOTES

[1] For an informative presentation and discussion of these results, see Edward Stein's Without Good Reason (Oxford: Oxford University Press (1996)).
[2] Cf. what Stein calls "the standard view" of rationality.
[3] Stein, op. cit.

L. JONATHAN COHEN

HOW CAN FALLACIES ARISE ABOUT FALLACIES?

1.

A hundred years ago, even perhaps as recently as fifty years ago, progress would normally be made in experimental psychology, in formal logic, in philosophical analysis, in linguistic theory or in neurology, as if those were essentially distinct and independent subjects of enquiry. Moreover this standard feature of intellectual research was unhesitatingly endorsed by the structure of library catalogues, of university departments, of funding agencies and of professional journals. Even computer science was treated largely as just an ancillary to military codebreaking. But all these areas ofresearch have now been absorbed and unified within the newly emerging ferment ofideas that has come to be known as cognitive science. So it is with the feeling ofbeing on the crest of a wave in current scientific enquiry that we meet here at the 1997 International Colloquium on Cognitive Science which has been so efficiently organised by our joint hosts, the Institute for Logic, Cognition, Language and Information and the Department of Logic and Philosophy of Science ofthe University ofthe Basque Country.

The issue that I myself wish to address is one that will very clearly illustrate, I hope, the value of this interdisciplinary approach. It concerns a special group of alleged fallacies that have recently come to the fore. No-one would want to dispute that people often commit fallacies, in the sense that they often reason invalidly. It is also beyond dispute that those who utter or hear such reasonings often fail to recognize that the reasonings are invalid. Much has accordingly been written during the past two and a half thousand years about what fallacies are particularly common, how fallacies should or should not be classified, when and where fallacies tend to occur, how detection of them is related to logical theory or to probability theory, and so on.[1] Until relatively recently, however, descriptions of how people actually reason were constructed observationally rather than experimentally. Fallacies were treated as types of event that occur frequently enough to be easily noticed in everyday life without any need to be provoked by experimental enquiry. Consequently little use was made ofexperimental methods in the attempt to enquire whether some, most, or all people are congenitally prone to reason invalidly in certain kinds of way. That type of experimental enquiry began to achieve prominence in the later 1960's and led to claims that psychological discoveries had been made which 'have bleak implications for human rationality'.[2] It was widely held then that certain types of systematic error, sometimes called 'heuristics and biases', were demonstrably common in untutored deductive and probabilistic reasoning, but were nevertheless erroneous. The existence of these heuristics and

K. Korta et al. (eds.), Cognition, Agency and Rationality, 9-19.

biases explained, it was supposed, many of the apparent errors of reasoning that are regularly made by subjects in the experiments.

But around 1980 most, if not all, of this psychological literature began to come under criticism from myself, Gerd Gigerenzer, and others.[3] To put it briefly we critics argued that the experimental data had been wrongly construed, and therefore wrongly explained, by the 'bleak implications' party. In particular we argued that, although the subjects of the experiments, like other people, may have made many accidental errors, they were not in fact involved in any sustained patterns of fallacious reasoning. Instead, we critics argued, it was really the 'bleak implications' party who, despite their scientific credentials, had involved themselves in sustained patterns of fallacious reasoning by supposing that their own interpretations of the questions put to the subjects were the only appropriate ones and that their own answers to those questions were therefore uniquely correct.

So, if such critics are right, an interesting question arises. How can it be that a considerable number of professionally trained experimental psychologists have all committed similar fallacies in interpreting their subjects' responses, and what is the nature of their mistake? If this question can be satisfactorily answered, the case against the 'bleak implications' psychologists is thereby strengthened. We shall have not only a description of these experimental psychologists' errors but also an explanation of their occurrence. It has to be kept in mind, however, that this explanation should not be an ad hoc one. That is to say, the covering principle invoked by the explanation of these errors should not be one that serves only the purpose in hand. The explanation will carry weight just so far as the principle has a strong and independent rationale. But if an adequate explanation is not forthcoming the case against the 'bleak implications' psychologists is thereby weakened: if their apparently fallacious reasonings have no regularly applicable explanation, then perhaps after all it is a mistake to suppose that their reasonings about the experimental data in question are regularly fallacious.

2.

I want to emphasise that my aim here is to put some of the facts of the matter within a unified overall perspective rather than to resolve each problem on its own. But let me begin in each case by reminding you of some of the relevant experimental data, the interpretations placed on them by the 'bleak implications' psychologists, and the reasons that can be given for rejecting those interpretations. Some of the experimental data concern judgements of probability and some concern judgements of deducibility. I shall deal first with some of the main issues about the former, and then with some of those about the latter. And in conclusion I shall draw attention to some of the main ways in which these two sets of issues are interconnected.

In one experiment[4] subjects were shown brief personality descriptions of several individuals who had allegedly been selected at random from a group of 100 professional people each of whom was either an engineer or a lawyer. In one trial of the experiment subjects were told that this group consisted of 70 engineers and 30 lawyers, in another that it consisted of 30 engineers and 70 lawyers. In each trial the ratio of the number of engineers to the total number of people involved was called the base-rate for that trial. All subjects were then asked by the experimenters to assess, for each description of a personality, the probability that it belonged to an engineer rather than to a lawyer; and in their answers the subjects tended to ignore the difference in base-rate between one trial and another. But the experimenters inferred that the desired probability should be higher in the first trial than in the second because they assumed that the data about the composition of the groups—that is, the 70% or 30% base-rate—supplied a prior probability in each case. Or in other words the experimenters applied the equation that is often called Bayes' theorem, viz. p(E) x p(H/E) = p(E/H) x p(H), and assumed that the prior probability p(H) was .7 in the one trial and .3 in the other, in each case putting the prior probability equal to the so-called 'base-rate'. Thus according to the experimenters most of the subjects were committing a fallacy, which they, the experimenters, called the 'base-rate fallacy'.

It is arguable, however, that the experimenters, not the subjects, were in error here. The experimenters assumed, by their application of Bayes' theorem, that they were using the same probability-metric throughout. In particular they assumed that in the context of their calculations the meaning of the term 'probability' in the sentence 'The probability of H is r' is the same as its meaning in 'The probability of H given E is r'. But in fact the meaning is different, because the term has different implications in the two contexts. From 'The probability of H given E is r' we can infer that, if George had been a member of the group (which he was not), then on the evidence of his description there could be just the same probability of his being an engineer as if he had in fact been a member of the group. But, so far as the probability is based on the ratio of the number of engineers to the number of individuals in the group as a whole, the probability might not be the same if the numbers in the group were altered by George's being a member of it. Or, in other words, assessment of a probability that stems from a personality description is unaffected by whether the individual involved is real or counterfactual, whereas assessment of a probability that stems from a relative frequency is subject to alteration if the number of members of the reference-class is changed. We have here two different types of probability-function, a counterfactualisable one and a non-counterfactualisable one, —a probability that is predicable of an individual on the basis of a personality description, and a probability that is predicable of a group of individuals on the basis of a relative frequency. So we must calculate accordingly. Either all the probabilities cited are to be counterfactualisable ones or all are to be non-counterfactualisable ones. And it is vital to distinguish between these two types of probability. The ambiguity must be resolved.

3.

A similar fallacy of undisambiguated probability occurs in what may be called the paradox of the taxi-cabs, where most people's pre-theoretical intuitions seem again to differ from what is alleged by some experimenters to be implied by elementary statistical theory.[5] Subjects are told by the experimenters that the vehicle involved in a certain road-accident in Smithville was a taxi-cab and that 85% of Smithville cabs are blue and 15% green. They are also told that a witness has identified the cab in the accident as green, and that this witness, when tested over equal numbers of blue and green cabs in conditions similar to those of the accident, gives the correct colour in 80% of cases and the other colour in only 20%. Subjects are then asked by experimenters what is the probability, expressed as a percentage, that the cab in the accident was blue. And the experimental evidence is that most statistically untutored subjects tend to estimate this probability as 20%. Thus the subjects seem to rely solely on the witness's testimony and to ignore altogether the specified base-rate— namely, the distribution of cab-colours. But, if the specified base-rate is treated as the relevant prior probability when the required posterior probability is calculated by means of Bayes' theorem, then this required probability amounts instead to approximately 60%. Accordingly the experimenters again condemn most of their subjects for committing here what they call the base-rate fallacy.

In fact, however, what happens in this type of experiment is that the experimenters assume that the ratio of accident-proneness among Smithville cabs under the relevant conditions, which is the prior probability needed for an application of Bayes' theorem, is the same as the ratio of blue cabs to green ones. And therein lies the experimenter's fallacy, because there are indeterminately many factors that might cause cabs of the rarer colour to be actually more accident-prone, under the relevant conditions, than cabs of the commoner colour. Or —to put the point another way— the experimenters are failing to distinguish between two different probabilities. One of these probabilities is counterfactualisable and is equatable with the ratio of accident-proneness among Smithville cabs under the relevant conditions, while the other is non-counterfactualisable and is equatable with the ratio of blue cabs in Smithville to blue ones. It is not the subjects who are committing a base-rate fallacy but the experimenters who are committing a fallacy of undisambiguated probability.

Nor is it difficult to conceive of any number of other cases in which such an issue— analogous to the Smithville one —might arise. For example, consider a clinical condition in which you can have only one or the other of only two diseases. One disease is found among 85% of the population, but a clinical test that has an 80% success record diagnoses the other disease. Which disease would you calculate to be the more probable one?

4.

In another experiment the data have been interpreted as showing a different type of irrationality. Instead of the alleged base-rate fallacy statistically untutored people are alleged to be unaware of the principle that larger samples of a given population show less variance. Subjects are instructed as follows[6]:

> A certain town is served by two hospitals. In the larger hospital about 45 babies are born each day, and in the smaller hospital about 15 babies are born each day. As you know, about 50% of all babies are boys. However, the exact percentage varies from day to day. Sometimes it may be higher than 50%, sometimes lower. For a period of one year each hospital recorded the days on which more than 60 of the babies born were boys. Which hospital do you think recorded more such days?

Because most subjects replied that both hospitals recorded about the same number of relevant days, and the other subjects were evenly divided between the larger number of days and the smaller, the experimenters concluded that most of their subjects were in error. They took this and similar results to demonstrate that the principle that sampling variance decreases in proportion to sample size 'is evidently not part ofpeople's repertoire of intuitions'. But that interpretation of the results assumes the facts given to relate —irrespective of all other matters— to statistical populations of baby-births in general, whereas the actual facts relate specifically to births that have occurred in particular hospitals. So the experimenters are thinking abstractly, in terms of idealised populations that have no other characteristic than their size, and are assuming idealised sampling from such populations, whereas the subjects, in accordance with their instructions, must bear in mind the concrete possibility —however remote— that the relative sizes of the two hospitals are causally connected with fluctuations in the boy-girl birth ratio. For example, perhaps both hospitals have a policy about allowing parents to choose the sex of their surviving children and perhaps this policy is more subject to political interference in the one hospital than in the other. Or in other words the experimenters are concerned with probabilities that are mathematical properties of sample-size in general, whereas most ofthe subjects are concerning themselves with probabilities that are rooted in facts about the causal ordering of events in society. Nor is there in practice any difficulty in cueing people (even children) to exhibit awareness ofthe principle that sampling variance decreases in proportion to sample size. That is to say, where it is appropriate to mathematicise about this issue, there is no difficulty in eliciting requisite intuitions from subjects.[7] But the experimenters are committing a fallacy of undisambiguated probability when they fail to allow for the two different ways —mathematical and factual— in which it is possible to approach the two-hospitals problem.

5.

Consider also what is implied when one person condemns another for having committed the so-called 'Gambler's Fallacy' or for believing in the so-called 'Maturity of Chances'. Whatever difficulties there may be in establishing the common occurrence of other kinds of fallacious probability judgement, it may be claimed that occurrence of the so-called 'Gambler's Fallacy' is surely well established. The Gambler here is a person who claims to be engaged in a game of pure chance but nevertheless holds that, the larger the number ofoutcomes since one particular type of outcome last occurred, the more probable it is that this type of outcome will occur next: its chance has 'matured'. A long run of heads, says the Gambler, makes it more probable than not that an outcome of tails will be next. And the Gambler's opponent would no doubt fortify his accusation of fallacy by pointing out that in a game of pure chance each outcome is totally independent of every other and neither learns from past outcomes nor is influenced by them in any way, so that one type of outcome does not become more probable than another by virtue of what actually happens on any occasion on which the game is actually played.

But the so-called 'Gambler's Fallacy' can be seen not to be a fallacy at all if it is recognised that there are other ways of conceiving a set of mutually exclusive probabilities than as the alternative outcomes in a game of pure chance. Thus in the real world of played games a long run of heads, for example, is normally taken to show a high probability that the coin, or the method of tossing it, is biased, not that there is a .5 probability of getting tails on the next toss. Similarly to find 13 hearts lieing successively in a pack of cards is normally taken to show a high probability that the pack has not been shuffled sufficiently, not that there is a .25 probability of being dealt a heart on the next deal. More precisely we can say, for example, that, if the results of the first ninety-nine tosses thrown so far have all been heads, there is a high probability that either the coin or the method of tossing is biased. Consequently, if we have to assume that, despite this probability, the game is one of pure chance and that neither the coin nor the method of tossing it is indeed biased, then there is a high probability that the results of the first 100 tosses so far are not all heads and thus a high probability that if the first ninety-nine tosses are indeed heads the hundredth will be tails.

What is happening here is that two different concepts of probability are in use. One of these is a relatively first-order concept of probability that is implicit in the description of a pattern of outcomes as being more or less probable in the context of some appropriate reference-class. And the other relevant concept of probability is a relatively second-order one, implicit in the description of anything as being probably an unbiased game, i.e. a game of pure chance, a game in which all the basic, mutually exclusive, outcomes are equiprobable. In short, without being able to use —implicitly or explicitly— these two different concepts of probability the Gambler lacks the conceptual resources to formulate his point of view coherently and becomes a prey to criticisms built on oversimplification. But with adequate disambiguation of the issues the criticisms are seen to be groundless.

6.

So far I have assumed that, though the semantics of a probability-function may vary from context to context, its syntax remains constant. Thus, while a judgement of probability may be either counterfactualisable or non-counterfactualisable, for example, and either a priori or a posteriori, the Pascalian calculus of chance supplies a permanent framework within which such judgements may be logically or mathematically related to one another. But, just as in addition to standard, Euclidean geometries there can also be non-standard, non-Euclidean geometries with distinctive axioms (about parallel lines, for example), so too in addition to standard, Pascalian probabilities, there can also be non-Pascalian probabilities —under whatever names— with distinctive axioms (about conjunction, negation, etc.). For example, it might be useful to explore the concept of weight, as Keynes called it,[8] and determine some of its applications,[9] or, if probability is thought of as strength of belief, one might think it useful to have a concept of probability in which it is possible to have both a weak degree of probability that p and a weak degree that not-p (which is, of course, impossible in Pascalian probability).

Indeed, some experimenters have committed the fallacy of undisambiguated probability in relation to an issue about conjunction. They claim that statistically untutored subjects often estimate the probability of a conjunction to be greater than that of one of the conjuncts, which can never be the case if our concept of probability is regulated by the mathematical calculus of chance. For example, if presented with Linda's profile in concrete detail, subjects may judge it rather improbable that she is a bank-teller, and rather more probable that she is a bank-teller and an active feminist.[10] So the experimenters are correct in attributing a fallacy to these subjects only if the subjects, when they interpret the experimenters' instructions, are indeed using a concept of probability that is regulated by the calculus of chance. But why should they be using such a concept then? After all, it was not until the middle of the seventeenth century that anyone thought of using the calculus of chance as a mathematics for measurements of evidential support.[11] So it is scarcely surprising that, while statistically sophisticated respondents immediately recognise the abstract validity of the conjunction rule, statistically naive respondents do not. Nor is the situation correctly described by saying that the statistically naive subjects are in error. It is not even that they lack an implicit awareness of how the abstract chance of a conjunction of outcomes relates to the abstract chances of the individual outcomes. In fact they can be shown to possess this awareness for use on appropriate occasions.[12] The only error that can legitimately be attributed to them is that of unwittingly indulging a gap in their education. They have not yet learned to use a concept of probability or of evidential support that is at least implicitly, if not explicitly, regulated by the calculus of chance. Indeed they may be supposed to assess the strength of the evidence that one proposition provides for another in terms of the concrete causal probability of such a connexion. If in the light of the evidence it is causally more coherent, and more characteristic of people with Linda's type of profile, for it to be supposed that Linda is a bank-teller and an active feminist than just that she is a bank-teller, it is quite reasonable for a statistically untutored subject to call the former conclusion 'more probable' than the latter.

Suppose you were a statistically untutored juror listening to one witness after another, then, other things being equal, wouldn't you give greater credence to the more concrete and coherent account, in which the conclusion was more representative of the evidence? Isn't that the account that you would think it more appropriate to believe?

7.

To sum up what has been established so far, we can say that what lies at the heart of the matter is the need to maintain a pluralistic stance in the conceptual analysis of probability. If we do maintain this, there is then no room for the accusations of irrationality that the psychologists cited have made against everyday assessments of probabilistic reasoning. Indeed we have here a clear example of the value of treating cognitive science as a unified, or potentially unifiable, field of research, as I characterised it at the beginning of the paper.

Of course, if the subjects can afterwards be brought to understand the questions in the way that the experimenters intend and to think of this interpretation as the correct one, they may even take some of their previous answers to have been mistaken. But at that stage they would have over-corrected. What the experimenters' critics have claimed is just that the subjects' initial answers show the meaning which they initially attribute to the questions and that these answers are not normally incorrect for those meanings.

Moreover an analogous pluralism is needed in relation to judgements of deducibility, as distinct from judgements of probability: after all, deducibility may be looked at as a limiting case of probability.[13] For example, during the last two thousand years or so logicians have often criticised the tendency of a logically naive reasoner to commit what they call the fallacy of illicit conversion. Such an allegedly fallacious reasoner infers or accepts a conclusion of the form 'If B, then A' from a premiss of the form 'If A, then B'. And the criticism of this inference is a valid one whenever the critics are entitled to assume that the elements functioning as premisses or conclusions here are sentences or propositions, which have no other relevant features than their truth-values. Where 'A', 'B', 'If A, then B', and 'If B, then A' stand in for abstract elements of such a kind, it is indeed a fallacy to suppose that 'If B, then A' is deducible from 'If A, then B'. But, if we take 'A', 'B', 'If A, then B, and 'If B, then A' to stand in for concrete utterances in specific contexts, we then have to bear in mind that the information provided by the utterance of such a sentence normally includes Gricean implicatures[14] stemming from the way in which human conversation is subject to rules of relevance, brevity, informativeness, and so on, as required by the purpose in hand. Unless there are indications to the contrary, this information may be presumed to be all that is required in the circumstances to satisfy the interest either of someone who wants to know what is also true if the antecedent of the conditional is true, or of someone who wants to know the conditions under which the consequent of the conditional sentence is true. In the former case (for example, 'If you interrupt him now, he'll be cross) the conditional is convertible because its utterance would normally be pointless unless 'If you don't interrupt him, he won't be cross' were also true. (Why bother to issue just the warning that if you interrupt him now he'll be cross, unless he won't be cross if you don't interrupt him?) In the other case (for example, 'If you give him a tip, he'll let you in') the conditional is convertible because its solitary utterance may be presumed to state what is the only condition under which the consequent is true.

Hence, if we consider the total concrete content of the message communicated rather than just the abstract conditional sentence that is uttered, it would not be fallacious or unwarranted for the subjects of a psychological experiment to presume, unless there are specific indications to the contrary, that the converse of an uttered conditional is implicit in the utterance. A psychological experimenter who wishes to exclude the legitimacy of presuming the converse in such a case must contrive suitable instructions to give to subjects and must teach them how to distinguish between the implications of a sentence and the implicatures of its utterance. If the experimenter does not do this (and most do not), he would himself commit one form of the fallacy of undisambiguated deducibility.

A somewhat similar result arises in any context in which it is normal for people who are posed logical conundrums to take into account relevant premisses that are already known to them. Thus an illiterate farmer may be told 'To carry corn from his farm, Bill needs a horse and a cart; he has the horse; but he does not have the cart.' The question is then put 'Can Bill carry corn from his farm?' The farmer replies 'Yes' on the grounds—quite reasonable in his culture—that Bill will borrow a cart. And to accuse such a person of transgressing the logical law of modus tollendo tollens would be a way of committing the fallacy of undisambiguated deducibility. It would assume an intention to construe the question in abstract logical termes, when the actual intention was to take the relevant concrete context into account (as psychologists have in effect recognised[15]).

Or consider the theorem of the propositional calculus which allows us to derive the formula '(A→D)∨(C→B)' from '(A→B)&(C→D)'. That derivation is a valid move in truth-functional logic, and yet it is not difficult to assign an ordinary-language interpretation to the derivation that makes this derivation highly paradoxical. For in ordinary language we could be deriving 'Either if John's automobile is a Mini, John is rich, or if John's automobile is a Rolls, John is poor' from 'If John's automobile is a Mini, John is poor, and if John's automobile is a Rolls, John is rich'. And this interpretation appears paradoxical because of the concrete context in which it is embedded—that is, because of what we know about the prices of certain cars. To ignore that embedding would be to commit the fallacy of undisambiguated deducibility. It would disregard the difference between deducibility as a function of formal derivability and deducibility as a function of readiness to believe.

8.

Another example of undisambiguated deducibility occurs in Wason's well-known four-card trick.[16] Subjects are presented with four laboratory cards showing, respectively, 'A', 'D', '4' and '7', and know from previous experience that every card, of which these are a subset, has a letter on one side and a number on the other. They are then given this rule about the four cards in front of them: 'If a card has a vowel on one side then it has an even number on the other'. Now they are told: 'Your task is to say which of the cards you need to turn over in order to find out whether the rule is true or false. The most frequent answers are 'A' and '4' and 'Only A', which are both wrong (and seem to indicate inability to operate properly with contrapositival reasoning), while the right answer 'A and 7' is given spontaneously by very few subjects.

If, however, real objects replace cards in the experiment, with a rule like 'If an envelope is sealed, then it has a fivepenny stamp on it' the subjects hardly ever come up with mistaken answers.[17] So it would be wrong to infer from the experiment that subjects are incapable of correct contrapositival reasoning. It is rather that logically unsophisticated subjects tend to operate with contrapositival reasoning in regard to concrete or causal cases (where the subjects are concerned with the behaviour of three-dimensional objects in the real world) and not in regard to abstract ones (where the subjects are concerned with artificially contrived situations and are insulated from most or all real-world parameters). Just why the subjects in the abstract cases make the particular mistakes that they do need not concern us here: it requires careful experimentation to determine. But the mistake that the experimenters originally made is clear enough. They thought that intelligent but logically unsophisticated subjects ought to be able to give the correct answers even in abstract cases, despite the fact that evolutionary pressures can be expected to select for concrete, causally sensitive abilities, which are usable in a range of familiar real-world problem situations, rather than for abstract, idealised ones, which are usable only under an artificially bounded range of unfamiliar laboratory conditions. And it did not occur at first to the experimenters that the cases in which subjects do tend to give correct answers might be the concrete ones and thus the ones in which evolutionary pressures can be expected to operate. In effect, therefore, the experimenters committed the fallacy of undisambiguated deducibility, because they did not allow for there to be a type of conditional rule with which the subjects can operate quite satisfactorily, as well as a type of conditional rule with which they cannot.

9.

My claim is then that the 'bleak implications' psychologists who mistakenly accuse their subjects of irrationality are in each case themselves guilty of a certain type of fallacy, namely, the fallacy of undisambiguated probability or undisambiguated deducibility. It is through committing that type of fallacy that certain professionally trained experimental scientists have come to make the mistakes that I have been discussing. But though this explains how such mistakes are possible, it leaves open the question why that type of fallacy should actually have been so common during the relevant period of psychological research. And the only suggestion that I can make about that question is that the operative cultural factor was a lack of appropriate communication between psychologists and philosophers. Philosophers were ready to envisage the existence of a plurality of types of probability or deducibility,[18] but experimental psychologists were not. Or perhaps one should rather say that it was easier to find sympathy for a pluralist approach to the analysis of cognitive tasks among philosophers than among psychologists. Correspondingly, now that within the scientific community we have moved towards the adoption of a more unified conception of cognitive science, we can expect this pluralist approach to become more widespread. That is to say, in the light of what philosophers have argued about probability and deducibility experimenters should expect to find signs of a plurality of concepts of probability and deducibility to be operative in their subjects' reasoning.

L. Jonathan Cohen
Oxford University
United Kingdom

NOTES

[1] For references see L.J. Cohen, The Dialogue of Reason, Oxford: Clarendon Press, 1986, p. 150ff.

[2] R.E. Nesbitt and E. Borgida, "Attribution and the psychology of prediction", *Journal of Personal and Social Psychology* 32, 1975, 932-943.

[3] See L.J. Cohen, "Can human irrationality be experimentally demonstrated?", *The Behavioral and Brain Sciences* 4, 1981, 317-370, and G. Gigerenzer, "How to make cognitive illusions disappear: beyond 'Heuristica and Biases'", in W. Stroebe and Miles Hewstone (eds.), *European Review of Social Psychology*, vol. 2, 1991, 83-115.

[4] A. Tversky, and D. Kahneman, "Judgement under uncertainty: heuristics and biases", *Science* 125, 1974, 1124-1131.

[5] A. Tversky, and D. Kahneman, "On the psychology of prediction", *Oregon Research Institute Research Bulletin* 12, 1972, 4.

[6] A. Tversky, and D. Kahneman, "Judgement under uncertainty: heuristics and biases", *Science* 125, 1974, 1124-1131.
[7] C.J. Jones, and P.L. Harris, "Insight into the law of large numbers: a comparison of Piagetian and judgement theory", *Quarterly Journal of Experimental Psychology,* 34A, 1982, 479-488.
[8] See L.J. Cohen, "Twelve questions about Keynes's concept of weight", *British Journal for the Philosophy of Science* 37, 1986, 263-278.
[9] See, for example, L.J. Cohen, *The Probable and the Provable*, Oxford: Clarendon Press, 1970.
[10] M. Bar-Hillel, "On the subjective probability of compound events", in *Organisational Behaviour and Human Performance* 9, 1973, 396-406.
[11] I. Hacking, The Emergence of Probability: a Philosophical Study of Early Ideas about Probability, Induction and Statistical Inference, Cambridge: Cambridge University Press.
[12] Transactions in Human Factors in Electronics HFE-7, 1966, 29-37; and L.R. Beach and C.R. Peterson, 'Subjective probabilities for unions of events', *Psychonomic Science* 5, 1966, 307-308.
[13] L.J. Cohen, "Probability —the one and the many", *Proceedings of the British Academy* LXI, 1975, 83-108.
[14] H.P. Grice, "Logic and conversation", in D. Davidson and G. Harman (eds.), *The Logic of Grammar,* Encino, California: Dickinson, 1975, 64-75.
[15] S. Scribner, "Modes of thinking and ways of speaking: culture and logic reconsidered", in P.N. Johnson-Laird and P.C. Wason (eds.), *Thinking; Readings in Cognitive Science,* Cambridge: Cambridge University Press, 1977, 483-500.
[16] P.C. Wason, "Reasoning", in B. Foss (ed.), *Psychology*, Harmondsworth: Penguin, New Horizons, 1966, 135-51.
[17] P.C. Wason, and D. Shapiro, "Natural and contrived experience in a reasoning problem", *Quarterly Journal of Experimental Psychology* 23, 1979, 63-71
[18] Notably R. Carnap, *Logical Foundations of Probability*, Chicago: Chicago University Press, 1950, 19-51. See also the references given in L.J. Cohen, *An Introduction to the Philosophy of Induction and Probability*, Oxford: Clarendon Press, 1989, 42.

RICHARD SAMUELS, STEPHEN STICH, PATRICE D. TREMOULET

RETHINKING RATIONALITY: FROM BLEAK IMPLICATIONS TO DARWINIAN MODULES

1. INTRODUCTION

There is a venerable philosophical tradition that views human beings as intrinsically rational, though even the most ardent defender of this view would admit that under certain circumstances people's decisions and thought processes can be very irrational indeed. When people are extremely tired, or drunk, or in the grip of rage, they sometimes reason and act in ways that no account of rationality would condone. About thirty years ago, Amos Tversky, Daniel Kahneman and a number of other psychologists began reporting findings suggesting much deeper problems with the traditional idea that human beings are intrinsically rational animals. What these studies demonstrated is that even under quite ordinary circumstances where fatigue, drugs and strong emotions are not factors, people reason and make judgments in ways that systematically violate familiar canons of rationality on a wide array of problems. Those first surprising studies sparked the growth of a major research tradition whose impact has been felt in economics, political theory, medicine and other areas far removed from cognitive science. In Section 2, we will sketch a few of the better known experimental findings in this area. We've chosen these particular findings because they will play a role at a later stage of the paper. For readers who would like a deeper and more systematic account of the fascinating and disquieting research on reasoning and judgment, there are now several excellent texts and anthologies available. (Nisbett and Ross 1980, Kahneman, Slovic and Tversky 1982, Baron 1988, Piatelli-Palmarini 1994, Dawes 1988, Sutherland 1994).

Though there is little doubt that most ofthe experimental results reported in the literature are robust and can be readily replicated, there is considerable debate over what these experiments indicate about the intrinsic rationality of ordinary people. One widely discussed interpretation of the results claims that they have "bleak implications" for the rationality of the man and woman in the street. What the studies show, according to this interpretation, is that ordinary people lack the underlying *competence* to handle a wide array of reasoning tasks, and thus that they must exploit a collection ofsimple *heuristics* which often lead to seriously counter-normative conclusions. Advocates of this interpretation would, of course, acknowledge that there are some people who have mastered the correct rules or procedures for handling some ofthese problems. But, they maintain, this knowledge

K. Korta et al. (eds.), Cognition, Agency and Rationality, 21-62.

is hard to acquire and hard to use. It is not the sort of knowledge that the human mind acquires readily or spontaneously in normal environments, and even those who have it often to do not use it unless they make a special effort. In Section 3, we will elaborate on this interpretation and explain the technical notion of competence that it invokes.

The pessimistic interpretation of the experimental findings has been challenged in a number of ways. One of the most recent and intriguing of these challenges comes from the emerging interdisciplinary field of evolutionary psychology. Evolutionary psychologists defend a highly *modular* conception of mental architecture which views the mind as composed of a large number of special purpose information processing organs or "modules" that have been shaped by natural selection to handle the sorts of recurrent information processing problems that confronted our hunter-gatherer forebears. Since good performance in a variety of reasoning tasks would likely have served our Pleistocene ancestors in good stead, evolutionary psychologists hypothesize that we should have evolved mental modules for handling these tasks well. However, they also maintain that the modules should be well adapted to the sorts of information that was available in the pre-human and early human environment. Thus, they hypothesize, when information is presented in the right way, performance on reasoning tasks should improve dramatically. In Section 4 we will offer a more detailed sketch of the richly modular picture of the mind advanced by evolutionary psychologists and of the notion of a mental module that plays a fundamental role in that picture. We will also take a brief look at the sorts of arguments offered by evolutionary psychologists for their contention that the mind is massively modular. Then, in Section 5, we will consider several recent studies that appear to confirm the evolutionary psychologists' prediction: When information is presented in ways that would have been important in our evolutionary history, performance on reasoning tasks soars. While the arguments and the experimental evidence offered by evolutionary psychologists are tantalizing, they hardly constitute a conclusive case for the evolutionary psychologists' theory about the mind and its origins. But a detailed critique of that theory would be beyond the scope of this paper. Rather, what we propose to do in our final section is to ask a hypothetical question. If the evolutionary psychologists' account turns out to be on the right track, what implications would this have for questions about the nature and the extent of human rationality or irrationality?

2. EXPLORING HUMAN REASONING AND JUDGMENT: FOUR EXAMPLES

2.1. The Selection Task

In 1966, Peter Wason reported the first experiments using a cluster of reasoning problems that came to be called the *Selection Task*. A recent textbook on reasoning has described that task as "the most intensively researched single problem in the history of the psychology of reasoning." (Evans, Newstead & Byrne 1993, p. 99.) A typical example of a Selection Task problem looks like this:

> Here are four cards. Each of them has a letter on one side and a number on the other side. Two of these cards are shown with the letter side up, and two with the number side up.
>
> | E | C | 5 | 4 |
> |---|---|---|---|
>
> Indicate which of these cards you have to turn over in order to determine whether the following claim is true:
>
> If a card has a vowel on one side, then it has an odd number on the other side.

What Wason and numerous other investigators have found is that subjects typically do very poorly on questions like this. Most subjects respond, correctly, that the E card must be turned over, but many also judge that the 5 card must be turned over, despite the fact that the 5 card could not falsify the claim no matter what is on the other side. Also, a large majority of subjects judge that the 4 card need *not* be turned over, though without turning it over there is no way of knowing whether it has a vowel on the other side. And, of course, if it does have a vowel on the other side then the claim is not true. It is not the case that subjects do poorly on all selection task problems, however. A wide range of variations on the basic pattern have been tried, and on some versions of the problem a much larger percentage of subjects answer correctly. These results form a bewildering pattern, since there is no obvious feature or cluster of features that separates versions on which subjects do well from those on which they do poorly. As we will see in Section 5, some evolutionary psychologists have argued that these results can be explained if we focus on the sorts of mental mechanisms that would have been crucial for reasoning about social exchange (or "reciprocal altruism") in the environment of our hominid forebears. The versions of the selection task we're good at, these theorists maintain, are just the ones that those mechanisms would have been designed to handle. But, as we will also see in Section 5, this explanation is hardly uncontroversial.

2.2. The Conjunction Fallacy

Ronald Reagan was elected President of the United States in November 1980. The following month, Amos Tversky and Daniel Kahneman administered a questionnaire to 93 subjects who had had no formal training in statistics. The instructions on the questionnaire were as follows:

> In this questionnaire you are asked to evaluate the probability of various events that may occur during 1981. Each problem includes four possible events. Your task is to rank order these events by probability, using 1 for the most probable event, 2 for the second, 3 for the third and 4 for the least probable event.

Here is one of the questions presented to the subjects:

> Please rank order the following events by their probability of occurrence in 1981:
>
> (a) Reagan will cut federal support to local government.
> (b) Reagan will provide federal support for unwed mothers.
> (c) Reagan will increase the defense budget by less than 5%.
> (d) Reagan will provide federal support for unwed mothers and cut federal support to local governments.

The unsettling outcome was that 68% of the subjects rated (d) as more probable than (b), despite the fact that (d) could not happen unless (b) did (Tversky & Kahneman 1982). In another experiment, which has since become quite famous, Tversky and Kahneman 1982 presented subjects with the following task:

> Linda is 31 years old, single, outspoken, and very bright. She majored in philosophy. As a student, she was deeply concerned with issues of discrimination and social justice, and also participated in anti-nuclear demonstrations.
>
> Please rank the following statements by their probability, using 1 for the most probable and 8 for the least probable.
>
> (a) Linda is a teacher in elementary school.
> (b) Linda works in a bookstore and takes Yoga classes.
> (c) Linda is active in the feminist movement.
> (d) Linda is a psychiatric social worker.
> (e) Linda is a member of the League of Women Voters.
> (f) Linda is a bank teller.
> (g) Linda is an insurance sales person.
> (h) Linda is a bank teller and is active in the feminist movement.

In a group of naive subjects with no background in probability and statistics, 89% judged that statement (h) was more probable than statement (f). When the same question was presented to statistically sophisticated subjects—graduate students in the decision science program of the Stanford Business School—85% made the same judgment! Results of this sort, in which subjects judge that a compound event or state of affairs is more probable than one of the components of the compound, have been found repeatedly since Kahneman and Tversky's pioneering studies.

2.3. Base-Rate Neglect

On the familiar Bayesian account, the probability of an hypothesis on a given body of evidence depends, in part, on the prior probability of the hypothesis. However, in a series of elegant experiments, Kahneman and Tversky 1973 showed that subjects often seriously undervalue the importance of prior probabilities. One of these experiments presented half of the subjects with the following "cover story."

> A panel of psychologists have interviewed and administered personality tests to 30 engineers and 70 lawyers, all successful in their respective fields. On the basis of this information, thumbnail descriptions of the 30 engineers and 70 lawyers have been written. You will find on your forms five descriptions, chosen at random from the 100 available descriptions. For each description, please indicate your probability that the person described is an engineer, on a scale from 0 to 100.

The other half of the subjects were presented with the same text, except the "base-rates" were reversed. They were told that the personality tests had been administered to 70 engineers and 30 lawyers. Some of the descriptions that were provided were designed to be compatible with the subjects' stereotypes of engineers, though not with their stereotypes of lawyers. Others were designed to fit the lawyer stereotype, but not the engineer stereotype. And one was intended to be quite neutral, giving subjects no information at all that would be of use in making their decision. Here are two examples, the first intended to sound like an engineer, the second intended to sound neutral:

> Jack is a 45-year-old man. He is married and has four children. He is generally conservative, careful and ambitious. He shows no interest in political and social issues and spends most of his free time on his many hobbies which include home carpentry, sailing, and mathematical puzzles.

> Dick is a 30-year-old man. He is married with no children. A man of high ability and high motivation, he promises to be quite successful in his field. He is well liked by his colleagues.

As expected, subjects in both groups thought that the probability that Jack is an engineer is quite high. Moreover, in what seems to be a clear violation of Bayesian principles, the difference in cover stories between the two groups of subjects had almost no effect at all. The neglect of base-rate information was even more striking in the case of Dick. That description was constructed to be totally uninformative with regard to Dick's profession. Thus the only useful information that subjects had was the base-rate information provided in the cover story. But that information was entirely ignored. The median probability estimate in both groups of subjects was 50%. Kahneman and Tversky's subjects were not, however, completely insensitive to base-rate information. Following the five descriptions on their form, subjects found the following "null" description:

> Suppose now that you are given no information whatsoever about an individual chosen at random from the sample.

> The probability that this man is one of the 30 engineers [or, for the other group of subjects, one of the 70 engineers] in the sample of 100 is ____%.

In this case subjects relied entirely on the base-rate; the median estimate was 30% for the first group of subjects and 70% for the second. In their discussion of these experiments, Nisbett and Ross offer this interpretation.

> The implication of this contrast between the "no information" and "totally nondiagnostic information" conditions seems clear. When *no* specific evidence about the target case is provided, prior probabilities are utilized appropriately; when *worthless* specific evidence is given, prior probabilities may be largely ignored, and people respond as if there were no basis for assuming differences in relative likelihoods. People's grasp of the relevance of base-rate information must be very weak if they could be distracted from using it by exposure to useless target case information. (Nisbett & Ross 1980, pp. 145-6.)

Before leaving the topic of base-rate neglect, we want to offer one further example illustrating the way in which the phenomenon might well have serious practical consequences. Here is a problem that Casscells et. al. 1978 presented to a group of faculty, staff and fourth-year students and Harvard Medical School.

> If a test to detect a disease whose prevalence is 1/1000 has a false positive rate of 5%, what is the chance that a person found to have a positive result actually has the disease, assuming that you know nothing about the person's symptoms or signs? ____%

Under the most plausible interpretation of the problem, the correct Bayesian answer is 2%. But only eighteen percent of the Harvard audience gave an answer close to 2%. Forty-five percent of this distinguished group completely ignored the base-rate information and said that the answer was 95%.

2.4. Over-Confidence

One of the most extensively investigated and most worrisome cluster of phenomena explored by psychologists interested in reasoning and judgment involves the degree of confidence that people have in their responses to factual questions—questions like:

> In each of the following pairs, which city has more inhabitants?
>
> (a) Las Vegas (b) Miami
> (a) Sydney (b) Melbourne
> (a) Hyderabad (b) Islamabad
> (a) Bonn (b) Heidelberg
>
> In each of the following pairs, which historical event happened first?
>
> (a) Signing of the Magna Carta (b) Birth of Mohammed
> (a) Death of Napoleon (b) Louisiana Purchase
> (a) Lincoln's assassination (b) Birth of Queen Victoria

After each answer subjects are also asked:

How confident are you that your answer is correct?

50% 60% 70% 80% 90% 100%

In an experiment using relatively hard questions it is typical to find that for the cases in which subjects say they are 100% confident, only about 80% of their answers are correct; for cases in which they say that they are 90% confident, only about 70% of their answers are correct; and for cases in which they say that they are 80% confident, only about 60% of their answers are correct. This tendency toward overconfidence seems to be very robust. Warning subjects that people are often overconfident has no significant effect, nor does offering them money (or bottles of French champagne) as a reward for accuracy. Moreover, the phenomenon has been demonstrated in a wide variety of subject populations including undergraduates, graduate students, physicians and even CIA analysts. (For a survey of the literature see Lichtenstein, Fischoff & Phillips 1982.)

3. BLEAK IMPLICATIONS: SHORTCOMINGS IN REASONING COMPETENCE

The experimental results we've been recounting and the many related results reported in the extensive literature in this area are, we think, intrinsically disquieting. They are even more alarming if, as has occasionally been demonstrated, the same patterns of reasoning and judgment are to be found outside the laboratory. None of us want our illnesses to be diagnosed by physicians who ignore well confirmed information about base-rates. Nor do we want our public officials to be advised by CIA analysts who are systematically overconfident. The experimental results themselves do not entail any conclusions about the nature or the normative status of the cognitive mechanisms that underlie people's reasoning and judgment. But a number of writers have urged that these results lend considerable support to a pessimistic hypothesis about those mechanisms, a hypothesis which may be even more disquieting than the results themselves. On this view, the examples of faulty reasoning and judgment that we've sketched are not mere *performance errors*. Rather, they indicate that most people's underlying *reasoning competence* is irrational or at least normatively problematic. In order to explain this view more clearly, we'll have to back up a bit and explain the rather technical distinction between competence and performance on which it is based.

The competence/performance distinction, as we will characterize it, was first introduced into cognitive science by Chomsky, who used it in his account of the explanatory strategy of theories in linguistics. (Chomsky 1965, Ch. 1; 1975; 1980) In testing linguistic theories, an important source of data are the "intuitions" or unreflective judgments that speakers of a language make about the grammaticality of sentences, and about various linguistic properties (e.g. Is the sentence ambiguous?) and relations (e.g. Is this phrase the subject of that verb?) To explain these intuitions, and also to explain how speakers go about producing and understanding sentences of their language in ordinary speech, Chomsky and his followers proposed what has become one of the most important hypotheses about the mind in the history of cognitive science. What this hypothesis claims is that a speaker of a language has an internally represented grammar of that language—an integrated set of generative rules and principles that entail an infinite number of claims about the language. For each of the infinite number of sentences in the speaker's language, the internally represented grammar entails that it is grammatical; for each ambiguous sentence in the speaker's language, the grammar entails that it is ambiguous, etc. When speakers make the judgments that we call linguistic intuitions, the information in the internally represented grammar is typically accessed and relied upon, though neither the process nor the internally represented grammar are accessible to consciousness. Since the internally represented grammar plays a central role in the production of linguistic intuitions, those intuitions can serve as an important source of data for linguists trying to specify what the rules and principles of the internally represented grammar are.

A speaker's intuitions are not, however, an infallible source of information about the grammar of the speaker's language, because the grammar cannot produce linguistic intuitions by itself. The production of intuitions is a complex process in which the internally represented grammar must interact with a variety of other cognitive mechanisms including those subserving perception, motivation, attention, short term memory and perhaps a host of others. In certain circumstances, the activity of any one of these mechanisms may result in a person offering a judgment about a sentence which does not accord with what the grammar actually entails about that sentence. The attention mechanism offers a clear example of this phenomenon. It is very likely the case that the grammar internally represented in typical English speakers entails that an infinite number of sentences of the form:

A told B that p, and B told C that q, and C told D that r, and ...

are grammatical in the speaker's language. However, if the present authors were asked to judge the grammaticality of a sentence containing a few hundred of these conjuncts, or perhaps even a few dozen, there is a good chance that our judgments would not reflect what our grammars entail, since in cases like this our attention easily wanders. Short term memory provides a more interesting example of the way in which a grammatical judgment may fail to reflect the information actually contained in the grammar. There is considerable evidence indicating that the short term memory mechanism has difficulty handling center embedded structures. Thus it may well be the case that our internally represented grammars entail that the following sentence is grammatical,

What what what he wanted cost would buy in Germany was amazing.

though our intuitions suggest, indeed shout, that it is not.

Now in the jargon that Chomsky introduced, the rules and principles of a speaker's internalized grammar constitutes the speaker's *linguistic competence*; the judgments a speaker makes about sentences, along with the sentences the speaker actually produces, are part of the speaker's *linguistic performance*. Moreover, as we have just seen, some of the sentences a speaker produces and some of the judgments the speaker makes about sentences, will not accurately reflect the speaker's linguistic competence. In these cases, the speaker is making a *performance error*.

There are some obvious analogies between the phenomena studied in linguistics and those studied by cognitive scientists interested in reasoning. In both cases there is spontaneous and largely unconscious processing of an open ended class of inputs; people are able to understand endlessly many sentences, and to draw inferences from endlessly many premises. Also, in both cases, people are able to make spontaneous intuitive judgments about an effectively infinite class of cases—judgments about grammaticality, ambiguity, etc. in the case of linguistics, and judgments about validity, probability, etc. in the case of reasoning. Given these analogies, it is plausible to explore the idea that the mechanism underlying our ability to reason is similar to the mechanism underlying our capacity to process language. And if Chomsky is right about language, then the analogous hypothesis about reasoning would claim that people have an internally represented integrated set of rules and principles of reasoning—a "psycho-logic" as it has been called—which is usually accessed and relied upon when people draw inferences or make judgments about them. As in the case of language, we would expect that neither the processes involved nor the principles of the internally represented psycho-logic are readily accessible to consciousness. We should also expect that people's inferences and judgments would not be an infallible guide to what the underlying psycho-logic actually entails about the validity or plausibility of a given inference. For here, as in the case oflanguage, the internally represented rules and principles must interact with lots of other cognitive mechanisms—including attention, motivation, short term memory and many others. The activity of these mechanisms can give rise to *performance errors*—inferences or judgments that do not reflect the psycho-logic which constitutes a person's *reasoning competence.*

There is, however, an important difference between reasoning and language, even if we assume that a Chomsky-style account of the underlying mechanism is correct in both cases. For in the case of language, it makes no clear sense to offer a normative assessment of a normal person's competence. The rules and principles that constitute a French speaker's linguistic competence are significantly different from the rules and principles that underlie language processing in a Chinese speaker. But if we were asked which system was better or which one was correct, we would have no idea what was being asked. Thus, on the language side of the analogy, there are performance errors, but there is no such thing as a competence error or a normatively problematic competence. If two otherwise normal people have different linguistic competences, then they simply speak different languages or different dialects. On the reasoning side ofthe analogy, things look very different. It is not clear whether there are significant individual and group differences in the rules and principles underlying people's performance on reasoning tasks, as there so clearly are in the rules and principles underlying people's linguistic performance. But if there are significant interpersonal differences in reasoning competence, it surely *appears* to make sense to ask whether one system ofrules and principles is better than another.[1] If Adam's psycho-logic ignores base-rates, endorses the conjunction fallacy and approves of affirming the consequent, while Bertha's does not, then, in these respects at least, it seems natural to say that Bertha's reasoning competence is better than Adam's. And even if all normal humans share the same psycho-logic, it still makes sense to ask how rational it is. If everyone's psycho-logic contains rules that get the wrong answer on certain versions of the selection task, then we might well conclude that there is a normative shortcoming that we all share.

We are now, finally, in a position to explain the pessimistic hypothesis that some authors have urged to account for the sort of experimental results sketched in Section 2. According to this hypothesis, the errors that subjects make in these experiments are very different from the sorts of reasoning errors that people make when their memory is overextended or when their attention wanders. They are also different from the errors people make when they are tired or drunk or blind with rage. These are all examples of *performance errors*—errors that people make when they infer in ways that are *not* sanctioned by their own psycho-logic. But the sorts of errors described in Section 2 are *competence errors*. In these cases people *are* reasoning and judging in ways that accord with their psycho-logic. The subjects in these experiments do not use the right rules because they do not have access to them; they are not part of the subjects' internally represented reasoning competence. What they have instead is a collection of simpler rules or "heuristics" that may often get the right answer, though it is also the case that often they do not. So according to this bleak hypothesis, the subjects make mistakes because their psycho-logic is normatively defective; their internalized rules of reasoning are less than fully rational. It is not at all clear that Kahneman and Tversky would endorse this interpretation of the experimental results, though a number of other leading researchers clearly do.[2] According to Slovic, Fischhoff and Lichtenstein, for example, "It appears that people lack the correct programs for many important judgmental tasks... We have not had the opportunity to evolve an intellect capable of dealing conceptually with uncertainty." 1976, p. 174)

Suppose it is in fact the case that many of the errors made in reasoning experiments are competence errors. That is not a flattering explanation, certainly, and it goes a long way toward undermining the traditional claim that man is a rational animal. But just how pessimistic a conclusion would it be? In part the answer depends on how hard would it be to improve people's performance, and that in turn depends on how hard it is to improve reasoning competence. Very little is known about that at present.[3] By invoking evolution as an explanation of our defective competence, however, Slovic, Fischhoff and Lichtenstein certainly do not encourage much optimism, since characteristics and limitations attributable to evolution are often innate, and innate limitations are not easy to overcome. The analogy with language points in much the same direction. For if Chomsky is right about language then, though it is obviously the case that people who speak different languages have internalized different grammars, the class of grammars that humans can internalize and incorporate into their language processing mechanism is severely restricted, and a significant part of an adult's linguistic competence is innate. If reasoning competence is similar to language competence, then it may well be the case that many improvements are simply not psychologically possible because our minds are not designed to reason well on these sorts of problems. This deeply pessimistic interpretation of the experimental results has been endorsed by a number of well-known authors, including Stephen J. Gould, who makes the point with his characteristic panache.

> I am particularly fond of [the Linda] example, because I know that the [conjunction] is least probable, yet a little homunculus in my head continues to jump up and down, shouting at me – "but she can't just be a bank teller; read the description." ... Why do we consistently make this simple logical error? Tversky and Kahneman argue, correctly I think, that our minds are not built (for whatever reason) to work by the rules of probability. (1992, p. 469.)

It is important to be clear about what it means to claim that improving our reasoning competence may be "psychologically impossible." In the case of language, people clearly do learn to use artificial languages like BASIC and LISP, which violate many of the constraints that a Chomskian would claim all natural (or "psychologically possible") languages must satisfy. However, people do not acquire and use BASIC in the way they acquire English or Arabic. Special effort and training is needed to learn it, and those who have mastered it only use it in special circumstances. No one "speaks" BASIC or uses it in the way that natural languages are used. Similarly, with special effort, it may be possible to learn rules of reasoning that violate some of the constraints on "natural" or "psychologically possible" rules, and to use those rules in special circumstances. But in confronting the myriad inferential challenges of everyday life, a person who had mastered a non-natural (but normatively superior) rule would typically use a less demanding and more natural "heuristic" rule. This is the point that Gould makes so vividly by conjuring a little homunculus jumping up and down in his head, and it might explain the otherwise surprising fact that graduate students in a prestigious decision science program are no better than the rest of us at avoiding the conjunction fallacy.

As we noted in the Introduction, there have been many attempts to challenge the pessimistic interpretation of the experimental findings on reasoning. In the two sections to follow we will focus on one of the boldest and most intriguing of these, the challenge from evolutionary psychology. If evolutionary psychologists are right, the rules and principles of reasoning available to ordinary people are much better than the "Bleak Implications" hypothesis would lead us to expect.

4. THE CHALLENGE FROM EVOLUTIONARY PSYCHOLOGY

In explaining the challenge from evolutionary psychology, the first order of business is to say what evolutionary psychology is, and that is not an easy task since this interdisciplinary field is too new to have developed any precise and widely agreed upon body of doctrines. There are, however, two basic ideas that are clearly central to evolutionary psychology. The first is that the mind consists of a large number of special purpose systems—often called "modules" or "mental organs." The second is that these systems, like other systems in the body, have been shaped by natural selection to perform specific functions or to solve information processing problems that were important in the environment in which our hominid ancestors evolved. In this section, we propose to proceed as follows. First, in 4.1, we'll take a brief look at some of the ways in which the notion of a "module" has been used in cognitive science, and focus in on the sorts of modules that evolutionary psychologists typically have in mind. In 4.2, we will contrast the massively modular account of the mind favored by evolutionary psychologists with another widely discussed conception of the mind according to which modules play only a peripheral role. In 4.3, we will consider an example of the sort of theoretical considerations that evolutionary psychologists have offered in support of their contention that the mind consists of large numbers of modules—and perhaps nothing else. Finally, in 4.4, we will give a very brief sketch of the evolutionary psychology research strategy.

4.1. *What Is a Mental Module?*

Though the term "module" has gained considerable currency in contemporary cognitive science, different theorists appear to use the term in importantly different ways. In this section we will outline some of these uses with the intention of getting a clearer picture of what evolutionary psychologists mean—and what they don't mean—by "module". The notions of modularity discussed in this section by no means exhausts the ways in which the term "module" is used in contemporary cognitive science. For a more comprehensive review see Segal 1996.

When speaking of modules, cognitive scientists are typically referring to mental structures or components of the mind that can be invoked in order to explain various cognitive capacities. Moreover, it is ordinarily assumed that modules are domain-specific (or functionally specific) as opposed to domain-general. Very roughly, this means that modules are dedicated to solving restricted classes of problems in unique domains. For instance, the claim that there is a vision module implies that there are mental structures which are brought into play in the domain of visual processing and are not recruited in dealing with other cognitive tasks. Later in this section we will discuss the notion of domain specificity in greater detail. For the moment, however, we want to focus on the fact that the term "module" is used to refer to two fundamentally different sorts of mental structures. (i) Sometimes it is used to refer to systems of mental representations. (ii) On other occasions the term "module" is used in order to talk about computational mechanisms. We will call modules of the first sort *Chomskian modules* and modules of the second sort *computational modules*.

4.1.1. *Chomskian Modules*

A Chomskian module is a domain specific body of mentally represented knowledge or information that accounts for a cognitive capacity. As the name suggests, the notion of a Chomskian module can be traced to Chomsky's work in linguistics. As we saw in Section 3, Chomsky claims that our linguistic competence consists in the possession of an internally represented grammar of our natural language. This grammar is a paradigm example of what we mean when speaking of Chomskian modules. But, of course, Chomsky is not the only theorist who posits the existence of what we are calling Chomskian modules. For instance, developmental psychologists such as Susan Carey and Elizabeth Spelke have argued that young children have domain-specific, mentally represented theories—systems of principles—for physics, psychology and mathematics (Carey and Spelke 1994). Theory-like structures of the sort posited by Carey and Spelke are an important kind of Chomskian module. However, if we assume that a theory is a *truth evaluable* system of representations, i.e. one in which it makes sense to ask whether the representations are true or false, then not all Chomskian modules must be theories. There can also be Chomskian modules that consist entirely of non-truth-evaluable systems of representations. There may, for example, be Chomskian modules that encode domain-specific knowledge of how to perform certain tasks—e.g. how to play chess, how to do deductive reasoning, or how to detect cheaters in social exchange settings.

As we have already noted, a domain-specific mental structure is one that is dedicated to solving problems in a restricted domain. In the case of Chomskian modules, it is ordinarily assumed that they are dedicated in this way for a specific reason: the content of the representations that constitute a given Chomskian module only represent properties and objects that belong to a specific domain. So, for example, if physics is a domain, then a Chomskian module for physics will only contain information about physical properties and physical objects. Similarly, if geometry constitutes a domain, then a Chomskian module for geometry will only contain information about geometrical properties and objects.

There are many problems with trying to characterize the notion of a Chomskian module in more precise terms. Clearly we do not want to treat just any domain specific collection of mental representations as a Chomskian module, since this would render the notion theoretically uninteresting. We do not, for example, want to treat a child's beliefs about toy dinosaurs as a module. Consequently, it is necessary to impose additional constraints in order to develop a useful notion of a Chomskian module. Two commonly invoked constraints are (i) innateness and (ii) restrictions on information flow. So, for example, according to Chomsky, Universal Grammar is an innate system of mental representations and most of the information that is contained in the Universal Grammar is not accessible to consciousness. (See Segal 1996 for an elaboration of these points.) We don't propose to pursue the issue of constraints any further, however, since as will soon become clear, when evolutionary psychologists speak of modules, they are usually concerned with a rather different kind of module—a computational module.

4.1.2. Computational Modules.

Computational modules are a species of computational device. As a first pass, we can characterize them as domain-specific, computational devices. A number of points of elaboration and clarification are in order, however. First, computational modules are ordinarily assumed to be classical computers, i.e. symbol (or representation) manipulating devices which receive representations as inputs and manipulate them according to formally specifiable rules in order to generate representations (or actions) as outputs. (For detailed discussions of the notion of classical computation see Haugeland 1985 and Pylyshyn 1984.) Classical computers of this sort contrast sharply with certain sorts of connectionist computational systems which cannot plausibly be viewed as symbol manipulating devices.[4]

Second, it is ordinarily assumed that computational modules are dedicated to solving problems in a specific domain because they are only capable of carrying out computations on a restricted range of inputs, namely representations of the properties and objects found in a particular domain. (Fodor 1983, p. 103.) So, for instance, if phonology constitutes a domain, then a phonology computational module will only provide analyses of inputs which are about phonological objects and properties. Similarly, if arithmetic is a domain, then an arithmetic computational module will only provide solutions to arithmetical problems.

Third, computational modules are usually assumed to be relatively autonomous components of the mind. Though they receive input from, and send output to, other cognitive processes or structures, they perform their own internal information processing unperturbed by external systems. For example, David Marr claims that the various computational modules on which parts of the visual process are implemented "are as nearly independent of each other as the overall task allows" (Marr 1982, p. 102).

Fourth, we want to emphasize the fact that computational modules are a very different kind of mental structure from Chomskian modules. Chomskian modules are *systems of representations*. By contrast, computational modules are processing devices—they *manipulate* representations. However, computational modules can co-exist with Chomskian modules. Indeed it may be that Chomskian modules, being bodies of information, are often manipulated by computational modules. Thus, for example, a parser might be conceived of as a computational module that deploys the contents of a Chomskian module devoted to linguistic information in order to generate syntactic and semantic representations of physical sentence-forms (Segal 1996, p. 144). Moreover, some Chomskian modules may be accessible only to a single computational module. When a Chomskian module and a computational module are linked in this way, it is natural to think of the two as unit, which we might call a *Chomskian/computational module*. But it is also important to note that the existence of Chomskian modules does not entail the existence of computational modules, since it is possible for a mind to contain Chomskian modules while not containing any computational modules. For example, while humans may possess domain-specific systems of knowledge for physics or geometry, it does not follow that we possess domain-specific computational mechanisms for processing information about physical objects or geometrical properties. Rather it may be that such domain-specific knowledge is utilized by domain-general reasoning systems.

A final point worth making is that the notion of a computational module has been elaborated in a variety of different ways in the cognitive science literature. Most notably, Fodor 1983 developed a conception of modules as domain-specific, computational mechanisms that are also (1) informationally encapsulated, (2) mandatory, (3) fast, (4) shallow, (5) neurally localized, (6) susceptible to characteristic breakdown, and (7) largely inaccessible to other processes.[5] Although the full fledged Fodorian notion of a module has been highly influential in cognitive science (Garfield 1987) evolutionary psychologists have not typically adopted his conception of modules. In his recent book, *Mindblindness*, for example, Simon Baron-Cohen explicitly denies that the modules involved in his theory of "mind reading"[6] need to be informationally encapsulated or have shallow outputs. (Baron-Cohen 1994, p. 515.)

4.1.3. Darwinian Modules.

What, then, do evolutionary psychologists typically mean by the term "module"? The answer, unfortunately, is far from clear, since evolutionary psychologists don't attempt to provide any precise characterization of modularity and rarely bother to distinguish between the various notions of module that we have set out in this section. Nevertheless, from what they do say about modularity, we think it is possible to piece together an account of what we propose to call a *Darwinian module*, which can be viewed as a sort of prototype of the evolutionary psychologists' notion of modularity. Darwinian modules have a cluster of features, and when evolutionary psychologists talk about modules they generally have in mind something that has most or all ofthe features in the cluster.

The first feature of Darwinian modules is that they are domain specific. According to Cosmides and Tooby, who are perhaps the best known proponents of evolutionary psychology, our minds consist primarily of "a constellation of specialized mechanisms that have domain-specific procedures, operate over domain-specific representations, or both. (Cosmides and Tooby 1994, p. 94.)

Second, Darwinian modules are computational mechanisms. On the colorful account offered by Tooby and Cosmides, "our cognitive architecture resembles a confederation of hundreds or thousands of functionally dedicated computers (often called modules) ..." (Tooby and Cosmides 1995, p. xiii.) Thus Darwinian modules are not Chomskian modules but rather a species of computational module. However, evolutionary psychologists also assume that many Darwinian modules utilize domain specific systems ofknowledge (i.e. Chomskian modules) when doing computations or solving problems, and that in some cases this domain specific knowledge is accessible only to a single Darwinian module. Thus some Darwinian modules are a kind of Chomskian/computational module. The "theory of mind" module posited by a number of recent theorists may provide an example. This module is typically assumed to employ innate, domain specific knowledge about psychological states when predicting the behavior of agents, and much of that information may not be available to other systems in the mind.

A third feature of Darwinian modules is that they are innate cognitive structures whose characteristic properties are largely or wholly determined by genetic factors. In addition, evolutionary psychologists make the stronger claim that the many Darwinian modules which predominate in our cognitive architecture are the products ofnatural selection. They are, according to Tooby and Cosmides, "kinds invented by natural selection during the species' evolutionary history to produce adaptive ends in the species natural environment." (Tooby and Cosmides 1995, p. xiii; see also Cosmides and Tooby 1992.) Thus, not only do evolutionary psychologists commit themselves to the claim that modules are innate, they also commit themselves to a theory about how modules came to be innate—viz. via natural selection. Though Darwinian modules need not enhance reproductive fitness in modern environments, they exist because they did enhance fitness in the environment ofour Pleistocene ancestors. Or, to make much the same point in the jargon favored by evolutionary psychologists, though Darwinian modules need not now be adaptive, they are *adaptations*. This account of the origins of these modules is, ofcourse, the reason that we have chosen to call them "Darwinian," and as we shall see in 4.4 the fact that Darwinian modules are adaptations plays an important role in structuring the research program that evolutionary psychologists pursue.

Finally, evolutionary psychologists often insist that Darwinian modules are universal features ofthe human mind and thus that we should expect to find that all (normally functioning) human beings possess the same specific set of modules. According to evolutionary psychologists, then, not only has natural selection designed the human mind so that it is rich in innate, domain-specific, computational mechanisms, but it has also given us all more-or-less the same design. (For an interesting critique of this claim, see Griffiths 1997, Ch. 5.)

To sum up, a (prototypical) Darwinian module is an innate, naturally selected, functionally specific and universal computational mechanism which may have access (perhaps even unique access) to a domain specific system of knowledge of the sort we've been calling a Chomskian module.

4.2. Peripheral Versus Massive Modularity

Until recently, even staunch proponents of modularity typically restricted themselves to the claim that the mind is modular at its periphery.[7] So, for example, although the discussion of modularity as it is currently framed in cognitive science derives largely from Jerry Fodor's arguments in *The Modularity of Mind* 1983, Fodor insists that much of our cognition is subserved by nonmodular systems. According to Fodor, only input systems (those responsible for perception and language processing) and output systems (those responsible for action) are plausible candidates for modularity. By contrast, "central systems" (those systems responsible for reasoning and belief fixation) are likely to be nonmodular. As Dan Sperber has observed:

> Although this was probably not intended and has not been much noticed, "modularity of mind" was a paradoxical title, for, according to Fodor, modularity is to be found only at the periphery of the mind... In its center and bulk, Fodor's mind is decidedly nonmodular. Conceptual processes —that is, thought proper— are presented as a holistic lump lacking joints at which to carve. (Sperber 1994, p. 39.)

Evolutionary psychologists reject the claim that the mind is only peripherally modular in favor of the view that the mind is largely or even entirely composed of Darwinian modules. We will call this thesis the Massive Modularity Hypothesis (MMH). Tooby and Cosmides elaborate on the Massive Modularity Hypothesis as follows:

> [O]ur cognitive architecture resembles a confederation of hundreds or thousands of functionally dedicated computers (often called modules) designed to solve adaptive problems endemic to our hunter-gatherer ancestors. Each of these devices has its own agenda and imposes its own exotic organization on different fragments of the world. There are specialized systems for grammar induction, for face recognition, for dead reckoning, for construing objects and for recognizing emotions from the face. There are mechanisms to detect animacy, eye direction, and cheating. There is a "theory of mind" module... a variety of social inference modules... and a multitude of other elegant machines. (Tooby and Cosmides 1995, p. xiv.)

According to MMH "central capacities too can be divided into domain-specific modules" (Jackendoff 1992, p. 70). So, for example, the linguist and cognitive neuroscientist Steven Pinker, has suggested that not only are there modules for perception, language and action, but there may also be modules for many tasks traditionally classified as central processes, including:

> Intuitive mechanics: knowledge of the motions, forces, and deformations that objects undergo... Intuitive biology: understanding how plants and animals work... Intuitive psychology: predicting other people's behavior from their beliefs and desires... [and] Self-concept: gathering and organizing information about one's value to other people, and packaging it for others. (Pinker 1994, p. 420.)

According to this view, then, "the human mind... [is]... not a general-purpose computer but a collection of instincts adapted for solving evolutionarily significant problems—the mind as a Swiss Army knife." (Pinker 1994.)[8]

4.3. Arguments For Massive Modularity

Is the Massive Modularity Hypothesis correct? Does the human mind consists largely or even entirely of Darwinian modules? This question that is fast becoming one of the central issues in contemporary cognitive science. Broadly speaking, the arguments in favor of MMH can be divided into two kinds, which we'll call "theoretical" and "empirical". Arguments of the first sort rely heavily on quite general theoretical claims about the nature of evolution, cognition and computation, while those of the second sort focus on experimental results which, it is argued, support the MMH view of the mind. While a systematic review of the arguments that have been offered in support of MMH would be beyond the scope of this essay, we think it is important for the reader to have some feel for what these arguments look like. Thus in this section we'll present a brief sketch of one of the theoretical arguments offered by Cosmides and Tooby, and suggest one way in which the argument might be criticized.[9] In Section 5, we'll consider some of the empirical results about reasoning that have been interpreted as supporting MMH.

Cosmides and Tooby's argument focuses on the notion of an *adaptive problem* which can be defined as an evolutionary recurrent problem whose solution promoted reproduction, however long or indirect the chain by which it did so (Cosmides and Tooby 1994, p. 87). For example, in order to reproduce, an organism must be able to find a mate. Thus finding a mate is an adaptive problem. Similarly, in order to reproduce, one must avoid being eaten by predators before one mates. Thus predator avoidance is also an adaptive problem. According to Cosmides and Tooby, once we appreciate both the way in which natural selection operates and the specific adaptive problems that human beings faced in the Pleistocene, we will see that there are good reasons for thinking that the mind contains a number of distinct, modular mechanisms. In developing the argument, Cosmides and Tooby first attempt to justify the claim that when it comes to solving adaptive problems, selection pressures can be expected to produce highly *specialized* cognitive mechanisms—i.e. modules.

> ... [D]ifferent adaptive problems often require different solutions and different solutions can, in most cases, be implemented only by different, functionally distinct mechanisms. Speed, reliability and efficiency can be engineered into specialized mechanisms because there is no need to engineer a compromise between different task demands. (Cosmides and Tooby 1994, p. 89.)

By contrast,

> ... a jack of all trades is necessarily a master of none, because generality can be achieved only by sacrificing effectiveness. (Cosmides and Tooby 1994, p. 89.)

In other words, while a specialized mechanism can be fast, reliable and efficient because it is dedicated to solving a specific adaptive problem, a general mechanism that solves many adaptive problems with competing task demands will only attain generality at the expense of sacrificing these virtues. Consequently:

> (1) As a rule, when two adaptive problems have solutions that are incompatible or simply different, a single solution will be inferior to two specialized solutions. (Cosmides and Tooby 1994, p. 89.)

Notice that the above quotation is not specifically about *cognitive* mechanisms. Rather it is supposed to apply generally to all solutions to adaptive problems. Nevertheless, according to Cosmides and Tooby, what applies generally to solutions to adaptive problems also applies to the specific case of cognitive mechanisms for solving adaptive problems. Thus, they claim, we have good reason to expect task specific or domain specific cognitive mechanisms to be superior solutions to adaptive problems than domain general systems. Moreover, since natural selection can be expected to favor superior solutions to adaptive problems over inferior ones, Cosmides and Tooby conclude that when it comes to solving adaptive problems:

> (2) ... domain-specific cognitive mechanisms ... can be expected to systematically outperform (and hence preclude or replace) more general mechanisms. (Cosmides and Tooby 1994, p. 89.)

So far, then, we have seen that Cosmides and Tooby argue for the claim that selection pressures can be expected to produce domain-specific cognitive mechanisms—modules—for solving adaptive problems. But this alone is not sufficient to support the claim that the mind contains a *large number* of modules. It must also be the case that our ancestors were confronted by a large number of adaptive problems that could be solved only by cognitive mechanisms. Accordingly, Cosmides and Tooby insist that

> (3) Simply to survive and reproduce, our Pleistocene ancestors had to be good at solving an enormously broad array of adaptive problems—problems that would defeat any modern artificial intelligence system. A small sampling include foraging for food, navigating, selecting a mate, parenting, engaging in social exchange, dealing with aggressive threat, avoiding predators, avoiding pathogenic contamination, avoiding naturally occurring plant toxins, avoiding incest and so on. (Cosmides and Tooby 1994, p. 90.)

Yet if this is true and if it is also true that when it comes to solving adaptive problems, domain-specific cognitive mechanisms can be expected to preclude or replace more general cognitive mechanisms, then it would seem to follow that:

> (4) The human mind can be expected to include a large number of distinct, domain-specific mechanisms.

And this, of course, is just what the Massive Modularity Hypothesis requires.

This argument is not supposed to be a deductive proof that the mind is massively modular. Rather it is offered as a plausibility argument. It is supposed to provide us with plausible grounds to expect the mind to contain many modules (Cosmides and Tooby 1994, p. 89). Nonetheless, if the conclusion of the argument is interpreted as claiming that the mind contains lots of *prototypical Darwinian* modules, then we suspect that the argument claims more than it is entitled to. For even if we grant that natural selection has contrived to provide the human mind with many specialized solutions to adaptive problems, it does not follow that these specialized solutions will be prototypical Darwinian modules. Rather than containing a large number of specialized computational devices, it might instead be the case that the mind contains lots of innate, domain specific items of knowledge, and that these are employed in order to solve various adaptive problems. Thus, rather than exploiting Darwinian modules, our minds might contain lots of innate, *Chomskian* modules. And it is perfectly consistent with the claim that we possess Chomskian modules for solving adaptive problems, that the information contained within such modules is utilized only by *domain-general* and, hence, nonmodular, computational devices. Moreover, the claim that natural selection prefers certain kinds of adaptive specializations to others—viz. Darwinian computational modules to Chomskian modules—surely does not follow from the general claim that specialized solutions (of some kind) typically outperform more general ones. So instead of producing Darwinian modules as solutions to adaptive problems, natural selection might instead have provided specialized solutions in the form of innate, domain-specific knowledge that is utilized by a domain-general computational mechanism. In order to make it plausible that the mind contains large numbers of Darwinian modules, one must argue for the claim that natural selection can be expected to prefer domain-specific *computational* devices over domain-specific *bodies of information* as solutions to adaptive problems. And, at present, it is far from clear that anyone knows how such an argument would go.

4.4 The Research Program of Evolutionary Psychology

A central goal of evolutionary psychology is to construct and test hypotheses about the Darwinian modules which, the theory maintains, make up much of the human mind. In pursuit of this goal, research may proceed in two quite different stages. The first, which we'll call *evolutionary analysis*, has as its goal the generation of plausible hypotheses about Darwinian modules. An evolutionary analysis tries to determine as much as possible about recurrent, information processing problems that our forebears would have confronted in what is often called *the environment of evolutionary adaptation* or the EEA—the environment in which *Homo Sapiens* evolved. The focus, of course, is on *adaptive* problems whose successful solution would have directly or indirectly contributed to reproductive success. In some cases these adaptive problems were posed by physical features of the EEA, in other cases they were posed by biological features, and in still other cases they were posed by the social environment in which our forebears were embedded. Since so many factors are involved in determining the sorts of recurrent information processing problems that our ancestors confronted in the EEA, this sort of evolutionary analysis is a highly interdisciplinary exercise. Clues can be found in many different sorts of investigations, from the study of the Pleistocene climate to the study of the social organization in the few remaining hunter-gatherer cultures. Once a recurrent adaptive problem has been characterized, the theorist may hypothesize that there is a module which would have done a good job at solving that problem in the EEA.

An important part of the effort to characterize these recurrent information processing problems is the specification of the sorts of constraints that a mechanism solving the problem could take for granted. If, for example, the important data needed to solve the problem was almost always presented in a specific format, then the mechanism need not be able to handle data presented in other ways. It could "assume" that the data would be presented in the typical format. Similarly, if it was important to be able to detect people or objects with a certain property that is not readily observable, and if, in the EEA, that property was highly correlated with some other property that is easier to detect, the system could simply assume that people or objects with the detectable property also had the one that was hard to observe.

It is important to keep in mind that evolutionary analyses can only be used as a way of *suggesting plausible hypotheses* about mental modules. By themselves evolutionary analyses provide no assurance that these hypotheses are true. The fact that it would have enhanced our ancestors' fitness if they had developed a module that solved a certain problem is no guarantee that they *did* develop such a module, since there are many reasons why natural selection and the other processes that drive evolution may fail to produce a mechanism that would enhance fitness. (Stich 1990, Ch. 3.)

Once an evolutionary analysis has succeeded in suggesting a plausible hypothesis, the next stage in the evolutionary psychology research strategy is to *test* the hypothesis by looking for evidence that contemporary humans actually have a module with the properties in question. Here, as earlier, the project is highly interdisciplinary. Evidence can come from experimental studies of reasoning in normal humans (Cosmides 1989, Cosmides and Tooby 1992, 1996, Gigerenzer 1991, Gigerenzer and Hug 1992), from developmental studies focused on the emergence of cognitive skills (Carey and Spelke 1994, Leslie 1994, Gelman and Brenneman 1994), or from the study of cognitive deficits in various abnormal populations (Baron-Cohen 1995). Important evidence can also be gleaned from studies in cognitive anthropology (Barkow 1992, Hutchins 1980), history, and even from such surprising areas as the comparative study of legal traditions (Wilson and Daly 1992). When evidence from a number of these areas points in the same direction, an increasingly strong case can be made for the existence of a module suggested by evolutionary analysis.

5. EVOLUTIONARY PSYCHOLOGY APPLIED TO REASONING: THEORY AND RESULTS

In this section we will consider two lines of research on human reasoning in which the two stage strategy described in the previous section has been pursued. Though the interpretation of the studies we will sketch is the subject of considerable controversy, a number of authors have suggested that they show there is something deeply mistaken about the "bleak" hypothesis set out in Section 3. That hypothesis claims that people lack normatively appropriate rules or principles for reasoning about problems like those set out in Section 2. But when we look at variations on these problems that may make them closer to the sort of recurrent problems our forebears would have confronted in the EEA, performance improves dramatically. And this, it is argued, is evidence for the existence of at least two normatively sophisticated Darwinian modules, one designed to deal with probabilistic reasoning when information is presented in a relative frequency format, the other designed to deal with reasoning about cheating in social exchange settings.

5.1. The Frequentist Hypothesis

The experiments reviewed in Sections 2.2-2.4 indicate that in many cases people are quite bad at reasoning about probabilities, and the pessimistic interpretation of these results claims that people use simple ("fast and dirty") heuristics in dealing with these problems because their cognitive systems have no access to more appropriate principles for reasoning about probabilities. But, in a series of recent and very provocative papers, Gigerenzer 1994, Gigerenzer & Hoffrage 1995 and Cosmides and Tooby 1996 argue that from an evolutionary point of view this would be a surprising and paradoxical result. "As long as chance has been loose in the world," Cosmides and Tooby note, "animals have had to make judgments under uncertainty." (Cosmides and Tooby 1996, p. 14; for the remainder of this section, all quotes are from Cosmides and Tooby 1996, unless otherwise indicated.) Thus making judgments when confronted with probabilistic information posed adaptive problems for all sorts of organisms, including our hominid ancestors, and "if an adaptive problem has endured for a long enough period and is important enough, then mechanisms ofconsiderable complexity can evolve to solve it." (p. 14.) But, as we saw in the previous section, "one should expect a mesh between the design of our cognitive mechanisms, the structure of the adaptive problems they evolved to solve, and the typical environments that they were designed to operate in—that is, the ones that they evolved in." (p. 14.) So in launching their evolutionary analysis Cosmides and Tooby's first step is to ask: "what kinds of probabilistic information would have been available to any inductive reasoning mechanisms that we might have evolved?" (p. 15.)

In the modern world we are confronted with statistical information presented in many ways: weather forecasts tell us the probability of rain tomorrow, sports pages list batting averages, and widely publicized studies tell us how much the risk of colon cancer is reduced in people over 50 if they have a diet high in fiber. But information about the probability of single events (like rain tomorrow) and information expressed in percentage terms would have been rare or unavailable in the EEA.

> What *was* available in the environment in which we evolved was the encountered frequencies of actual events—for example, that we were successful 5 times out of the last 20 times we hunted in the north canyon. Our hominid ancestors were immersed in a rich flow of observable frequencies that could be used to improve decision-making, given procedures that could take advantage of them. So if we have adaptations for inductive reasoning, they should take frequency information as input. (pp. 15-16.)

After a cognitive system has registered information about relative frequencies it might convert this information to some other format. If, for example, the system has noted that 5 out of the last 20 north canyon hunts were successful, it might infer and store the conclusion that there is a .25 chance that a north canyon hunt will be successful. However, Cosmides and Tooby argue, "there are advantages to storing and operating on frequentist representations because they preserve important information that would be lost by conversion to single-event probability. For example, ... the number of events that the judgment was based on would be lost in conversion. When the *n* disappears, the index of reliability of the information disappears as well." (p. 16.)

These and other considerations about the environment in which our cognitive systems evolved lead Cosmides and Tooby to hypothesize that our ancestors "evolved mechanisms that took frequencies as input, maintained such information as frequentist representations, and used these frequentist representations as a database for effective inductive reasoning."[10] Since evolutionary psychologists expect the mind to contain many specialized modules, Cosmides and Tooby are prepared to find other modules involved in inductive reasoning that work in other ways.

> We are not hypothesizing that every cognitive mechanism involving statistical induction necessarily operates on frequentist principles, only that at least one of them does, and that this makes frequentist principles an important feature of how humans intuitively engage the statistical dimension of the world. (p. 17.)

But, while their evolutionary analysis does not preclude the existence of inductive mechanisms that are not focused on frequencies, it does suggest that when a mechanism that operates on frequentist principles is engaged, it will do a good job, and thus the probabilistic inferences it makes will generally be normatively appropriate ones. This, of course, is in stark contrast to the bleak implications hypothesis which claims that people simply do not have access to normatively appropriate strategies in this area.

From their hypothesis, Cosmides and Tooby derive a number of predictions:

> (1) Inductive reasoning performance will differ depending on whether subjects are asked to judge a frequency or the probability of a single event.
>
> (2) Performance on frequentist versions of problems will be superior to non-frequentist versions.
>
> (3) The more subjects can be mobilized to form a frequentist representation, the better performance will be.
>
> (4) ... Performance on frequentist problems will satisfy some of the constraints that a calculus of probability specifies, such as Bayes' rule. This would occur because some inductive reasoning mechanisms in our cognitive architecture embody aspects of a calculus of probability. (p. 17.)

To test these predictions Cosmides and Tooby ran an array of experiments designed around the medical diagnosis problem which Casscells et al. used to demonstrate that even very sophisticated subjects ignore information about base rates. In their first experiment Cosmides and Tooby replicated the results of Casscells et al. using exactly the same wording that we reported in Sec. 2.4. Of the 25 Stanford University undergraduates who were subjects in this experiment, only 3 (= 12%) gave the normatively appropriate bayesian answer of "2%", while 14 subjects (= 56%) answered "95%".[11] As we noted in 2.3, the Harvard Medical School subjects in the original Casscells et al. study did slightly better; 18% of those subjects gave answers close to "2%" and 45% answered "95%".

In another experiment, Cosmides and Tooby gave 50 Stanford students a similar problem in which relative frequencies rather than percentages and single event probabilities were emphasized. The "frequentist" version of the problem read as follows:

> 1 out of every 1000 Americans has disease X. A test has been developed to detect when a person has disease X. Every time the test is given to a person who has the disease, the test comes out positive. But sometimes the test also comes out positive when it is given to a person who is completely healthy. Specifically, out of every 1000 people who are perfectly healthy, 50 of them test positive for the disease.
>
> Imagine that we have assembled a random sample of 1000 Americans. They were selected by lottery. Those who conducted the lottery had no information about the health status of any of these people.
>
> Given the information above:
>
> on average,
>
> How many people who test positive for the disease will *actually* have the disease? _____ out of _____.[12]

On this problem the results were dramatically different. 38 of the 50 subjects (= 76%) gave the correct bayesian answer.[13]

A series of further experiments systematically explored the differences between the problem used by Casscells et al. and the problems on which subjects perform well, in an effort to determine which factors had the largest effect. Although a number of different factors affect performance, two predominate. "Asking for the answer as a frequency produces the largest effect, followed closely by presenting the problem information as frequencies." (p. 58.) The most important conclusion that Cosmides and Tooby want to draw from these experiments is that "frequentist representations activate mechanisms that produce bayesian reasoning, and that this is what accounts for the very high level of bayesian performance elicited by the pure frequentist problems that we tested." (p. 59.)

As further support for this conclusion, Cosmides and Tooby cite several striking results reported by other investigators. In one study, Fiedler 1988, following up on some intriguing findings in Tversky and Kahneman 1983, showed that the percentage of subjects who commit the conjunction fallacy can be radically reduced if the problem is cast in frequentist terms. In the "feminist bank teller" example, Fiedler contrasted the wording reported in 2.2 with a problem that read as follows:

> Linda is 31 years old, single, outspoken, and very bright. She majored in philosophy. As a student, she was deeply concerned with issues of discrimination and social justice, and also participated in anti-nuclear demonstrations.

There are 200 people who fit the description above. How many of them are:

bank tellers?

bank tellers and active in the feminist movement?

...

In Fiedler's replication using the original formulation of the problem, 91% of subjects judged the feminist bank teller option to be more probable than the bank teller option. However in the frequentist version only 22% of subjects judged that there would be more feminist bank tellers than bank tellers. In yet another experiment, Hertwig and Gigerenzer 1994; reported in Gigerenzer 1994 told subjects that there were 200 women fitting the "Linda" description, and asked them to estimate the number who were bank tellers, feminist bank tellers, and feminists. Only 13% committed the conjunction fallacy.

Studies on over-confidence have also been marshaled in support of the frequentist hypothesis. In one of these Gigerenzer, Hoffrage and Kleinbölting 1991 reported that the sort of over-confidence described in 2.4 can be made to "disappear" by having subjects answer questions formulated in terms of frequencies. Gigerenzer and his colleagues gave subjects lists of 50 questions similar to those described in 2.4, except that in addition to being asked to rate their confidence after each response (which, in effect, asks them to judge the probability of that single event), subjects were, at the end, also asked a question about the frequency of correct responses: "How many of these 50 questions do you think you got right?" In two experiments, the average over-confidence was about 15%, when single-event confidences were compared with actual relative frequencies of correct answers, replicating the sorts of findings we sketched in Section 2.4. However, comparing the subjects' "estimated frequencies with actual frequencies of correct answers made 'overconfidence' *disappear*... Estimated frequencies were practically identical with actual frequencies, with even a small tendency towards underestimation. The 'cognitive illusion' was gone." (Gigerenzer 1991, p. 89.)

Both the experimental studies we have been reviewing and the conclusions that Gigerenzer, Cosmides, and Tooby want to draw from them have provoked a fair measure of criticism. For our purposes, perhaps the most troublesome criticisms are those demonstrating that various normatively problematic patterns of reasoning arise even when a problem is stated in terms of frequencies. In their detailed study of the conjunction fallacy, for example, Tversky and Kahneman 1983 reported an experiment in which subjects were asked to estimate both the number of "seven-letter words of the form '-----n-' in four pages of text" and the number of "seven letter words of the form '----ing' in four pages of text." The median estimate for words ending in "ing" was about three times *higher* than for words with "n" in the next-to-last position. As Kahneman and Tversky 1996 note, this appears to be a clear counter-example to Gigerenzer's claim that the conjunction fallacy disappears in judgments of frequency.

As another challenge to the claim that frequency representations eliminate base-rate neglect, Kahneman and Tversky cite a study by Gluck and Bower 1988). In that study subjects were required to learn to diagnose whether a patient had a rare disease (25%) or a common disease (75%) on the basis of 250 trials in which they were presented with patterns of 4 symptoms. After each presentation subjects guessed which disease the patient had, and were given immediate feedback indicating whether their guess was right or wrong. Though subjects encountered the common disease three times more often than the rare disease, they largely ignored this base rate information, and acted as if the two diseased were equally likely.

There is also a substantial body of work demonstrating that antecedent expectations can lead people to report illusory correlations when they are shown data about a sequence of cases. In one well known and very disquieting study, Chapman and Chapman 1967, 1969 showed subjects a series of cards each of which was said to reproduce a drawing of a person made by a psychiatric patient. Each card also gave the diagnosis for that patient. Subjects reported seeing "intuitively expected" correlations (e.g. drawings with peculiar eyes and diagnoses of paranoia) even when the there was no such correlation in the data they were shown. In another widely discussed study, Gilovich, Vallone and Tversky 1985 showed that people "see" a positive correlation between the outcome of successive shots in basketball (thus giving rise to the illusion of a "hot hand") even when there is no such correlation in the data.

On our view, what these criticisms show is that the version of the frequentist hypothesis suggested by Gigerenzer, Cosmides and Tooby is too simplistic. It is not the case that all frequentist representations activate mechanisms that produce good bayesian reasoning, nor is it the case that presenting data in a sequential format from which frequency distribution can readily be extracted always activates mechanisms that do a good job at detecting correlations. More experimental work will be needed to determine what additional factors are required to trigger good bayesian reasoning and good correlation detection. And more subtle evolutionary analyses will be needed to throw light on why these more complex triggers evolved. But despite the polemical fireworks, there is actually a fair amount of agreement between the evolutionary psychologists and their critics. Both sides agree that people *do* have mental mechanisms which can do a good job at bayesian reasoning, and that presenting problems in a way that makes frequency information salient can play an important role in activating these mechanisms. Both sides also agree that people have other mental mechanisms that exploit quite different reasoning strategies, though there is little agreement on how to characterize these non-bayesian strategies, what factors trigger them, or why they evolved. The bottom line, we think, is that the experiments demonstrating that people sometimes do an excellent job of bayesian reasoning go a long way toward refuting the gloomy hypothesis sketched in Section 3. Gould's claim that "our minds are not built ... to work by the rules of probability" is much too pessimistic. Our cognitive systems clearly do have access to reasoning strategies that accord with the rules of probability, though it is also clear that we don't always use them. We also think that the evidence reviewed in this section is compatible with the hypothesis that good probabilistic reasoning, when it occurs, is subserved by one or more Darwinian modules, though of course the evidence is compatible with lots of alternative hypothesis as well.

5.2. The Cheater Detection Hypothesis

In Section 2 we reproduced one version of Wason's four card selection task on which most subjects perform very poorly, and we noted that, while subjects do equally poorly on many other versions of the selection task, there are some versions on which performance improves dramatically. Here is an example from Griggs and Cox 1982:

> In its crackdown against drunk drivers, Massachusetts law enforcement officials are revoking liquor licenses left and right. You are a bouncer in a Boston bar, and you'll loose your job unless you enforce the following law:
>
> **"If a person is drinking beer, then he must be over 20 years old."**
>
> The cards below have information about four people sitting at a table in your bar. Each card represents one person. One side of a card tells what a person is drinking and the other side of the card tells that person's age. Indicate only those card(s) you definitely need to turn over to see if any of these people are breaking the law .
>
> | DRINKING BEER | DRINKING COKE | 25 YEARS OLD | 16 YEARS OLD |
> |---|---|---|---|

From a logical point of view this problem is structurally identical to the problem in Section 2.1, but the *content* of the problems clearly has a major effect on how well people perform. About 75% of college student subjects get the right answer on this version of the selection task, while only 25% get the right answer on the other version. Though there have been dozens of studies exploring this "content effect" in the selection task, the results have been, and continue to be, rather puzzling since there is no obvious property or set of properties shared by those versions of the task on which people perform well. However, in several recent and widely discussed papers, Cosmides and Tooby have argued that an evolutionary analysis enables us to see a surprising pattern in these otherwise bewildering results. (Cosmides 1989, Cosmides and Tooby 1992.)

The starting point of their evolutionary analysis is the observation that in the environment in which our ancestors evolved (and in the modern world as well) it is often the case that unrelated individuals can engage in "non-zero-sum" exchanges, in which the benefits to the recipient (measured in terms of reproductive fitness) are significantly greater than the costs to the donor. In a hunter-gatherer society, for example, it will sometimes happen that one hunter has been lucky on a particular day and has an abundance of food, while another hunter has been unlucky and is near starvation. If the successful hunter gives some of his meat to the unsuccessful hunter rather than gorging on it himself, this may have a small negative effect on the donor's fitness since the extra bit of body fat that he might add could prove useful in the future, but the benefit to the recipient will be much greater. Still, there is *some* cost to the donor; he would be slightly better off if he didn't help unrelated individuals. Despite this it is clear that people sometimes do help non-kin, and there is evidence to suggest that non-human primates (and even vampire bats) do so as well. On first blush, this sort of "altruism" seems to pose an evolutionary puzzle, since if a gene which made an organism *less* likely to help unrelated individuals appeared in a population, those with the gene would be slightly *more* fit, and thus the gene would gradually spread through the population.

A solution to this puzzle was proposed by Robert Trivers 1971 who noted that, while one-way altruism might be a bad idea from an evolutionary point of view, *reciprocal altruism* is quite a different matter. If a pair of hunters (be they humans or bats) can each count on the other to help when one has an abundance of food and the other has none, then they may both be better off in the long run. Thus organisms with a gene or a suite of genes that inclines them to engage in reciprocal exchanges with non-kin (or "social exchanges" as they are sometimes called) would be more fit than members of the same species without those genes. But of course, reciprocal exchange arrangements are vulnerable to cheating. In the business of maximizing fitness, individuals will do best if they are regularly offered and accept help when they need it, but never reciprocate when others need help. This suggests that if stable social exchange arrangements are to exist, the organisms involved must have cognitive mechanisms that enable them to detect cheaters, and to avoid helping them in the future. And since humans apparently are capable of entering into stable social exchange relations, this evolutionary analysis leads Cosmides and Tooby to hypothesize that we have one or more Darwinian modules whose job is to recognize reciprocal exchange arrangements and to detect cheaters who accept the benefits in such arrangements but do not pay the costs. In short, the evolutionary analysis leads Cosmides and Tooby to hypothesize the existence of one or more cheater detection modules. We call this *the cheater detection hypothesis*.

If this is right, then we should be able to find some evidence for the existence of these modules in the thinking of contemporary humans. It is here that the selection task enters the picture. For according to Cosmides and Tooby, some versions of the selection task engage the mental module(s) which were designed to detect cheaters in social exchange situations. And since these mental modules can be expected to do their job efficiently and accurately, people do well on those versions of the selection task. Other versions of the task do not trigger the social exchange and cheater detection modules. Since we have no mental modules that were designed to deal with these problems, people find them much harder, and their performance is much worse. The bouncer-in-the-Boston-bar problem presented earlier is an example of a selection task that triggers the cheater detection mechanism. The problem involving vowels and odd numbers presented in Section 2 is an example of a selection task that does not trigger cheater detection module.

In support of their theory, Cosmides and Tooby assemble an impressive body of evidence. To begin, they note that the cheater detection hypothesis claims that social exchanges, or "social contracts" will trigger good performance on selection tasks, and this enables us to see a clear pattern in the otherwise confusing experimental literature that had grown up before their hypothesis was formulated.

> When we began this research in 1983, the literature on the Wason selection task was full of reports of a wide variety of content effects, and there was no satisfying theory or empirical generalization that could account for these effects. When we categorized these content effects according to whether they conformed to social contracts, a striking pattern emerged. Robust and replicable content effects were found only for rules that related terms that are recognizable as benefits and cost/requirements in the format of a standard social contract... No thematic rule that was not a social contract had ever produced a content effect that was both robust and replicable... All told, for non-social contract thematic problems, 3 experiments had produced a substantial content effect, 2 had produced a weak content effect, and 14 had produced no content effect at all. The few effects that were found did not replicate. In contrast, 16 out of 16 experiments that fit the criteria for standard social contracts ... elicited substantial content effects. (Cosmides and Tooby 1992, p. 183.)

Since the formulation of the cheater detection hypothesis, a number of additional experiments have been designed to test the hypothesis and rule out alternatives. Among the most persuasive of these are a series of experiments by Gigerenzer and Hug 1992. In one set of experiments, these authors set out to show that, contrary to an earlier proposal by Cosmides and Tooby, *merely* perceiving a rule as a social contract was not enough to engage the cognitive mechanism that leads to good performance in the selection task, and that cueing for the possibility of *cheating* was required. To do this they created two quite different context stories for social contract rules. One of the stories required subjects to attend to the possibility of cheating, while in the other story cheating was not relevant. Among the rules social contract rules they used was the following which, they note, is widely known among hikers in the Alps:

> (i.) If someone stays overnight in the cabin, then that person must bring along a bundle of wood from the valley.

The first context story, which the investigators call the "cheating version," explained:

> There is a cabin at high altitude in the Swiss Alps, which serves hikers as an overnight shelter. Since it is cold and firewood is not otherwise available at that altitude, the rule is that each hiker who stays overnight has to carry along his/her own share of wood. There are rumors that the rule is not always followed. The subjects were cued into the perspective of a guard who checks whether any one of four hikers has violated the rule. The four hikers were represented by four cards that read "stays overnight in the cabin", "carried no wood", "carried wood", and "does not stay overnight in the cabin".

The other context story, the "no cheating version,"

> cued subjects into the perspective of a member of the German Alpine Association who visits the Swiss cabin and tries to discover how the local Swiss Alpine Club runs this cabin. He observes people bringing wood to the cabin, and a friend suggests the familiar overnight rule as an

> explanation. The context story also mentions an alternative explanation: rather than the hikers, the members of the Swiss Alpine Club, who do not stay overnight, might carry the wood. The task of the subject was to check four persons (the same four cards) in order to find out whether anyone had violated the overnight rule suggested by the friend. (Gigerenzer and Hug 1992, pp. 142-143.)

The cheater detection hypothesis predicts that subjects will do better on the cheating version than on the no cheating version, and that prediction was confirmed. In the cheating version, 89% of the subjects got the right answer, while in the no cheating version, only 53% responded correctly.

In another set of experiments, Gigerenzer and Hug showed that when social contract rules make cheating on both sides possible, cueing subjects into the perspective of one party or the other can have a dramatic effect on performance in selection task problems. One of the rules they used that allows the possibility of bilateral cheating was:

> (ii.) If an employee works on the weekend, then that person gets a day off during the week.

Here again, two different context stories were constructed, one of which was designed to get subjects to take the perspective of the employee, while the other was designed to get subjects to take the perspective of the employer.

> The employee version stated that working on the weekend is a benefit for the employer, because the firm can make use of its machines and be more flexible. Working on the weekend, on the other hand is a cost for the employee. The context story was about an employee who had never worked on the weekend before, but who is considering working on Saturdays from time to time, since having a day off during the week is a benefit that outweighs the costs of working on Saturday. There are rumors that the rule has been violated before. The subject's task was to check information about four colleagues to see whether the rule has been violated. The four cards read: "worked on the weekend", "did not get a day off", "did not work on the weekend", "did get a day off".
>
> In the employer version, the same rationale was given. The subject was cued into the perspective of the employer, who suspects that the rule has been violated before. The subjects' task was the same as in the other perspective [viz. to check information about four employees to see whether the rule has been violated]. (Gigerenzer & Hug 1992, p. 154.)

In these experiments about 75% of the subjects cued to the employee's perspective chose the first two cards ("worked on the weekend" and "did not get a day off") while less than 5% chose the other two cards. The results for subjects cued to the employer's perspective were radically different. Over 60% of subjects selected the last two cards ("did not work on the weekend" and "did get a day off") while less than 10% selected the first two.

The evolutionary analysis that motivates the cheater detection hypothesis maintains that the capacity to engage in social exchange could not have evolved unless the individuals involved had some mechanism for detecting cheaters. There would, however, be no need for our hominid forebears to have developed a mechanism for detecting "pure altruists" who help others but do not expect help in return. If there were individuals like that, it might of course be useful to recognize them so that they could be more readily exploited. However, altruists of this sort would incur fitness costs with no compensating benefits, and thus an evolutionary analysis suggests that they would have been selected against. Since altruists would be rare or non-existent, there would be no selection pressure for an altruist detection mechanism. These considerations led Cosmides and Tooby to predict that people will be much better at detecting cheaters in a selection task than at detecting altruists. To test the prediction they designed three pairs of problems. In each pair the two stories are quite similar, though in one version subjects must look for cheaters, while in the other they must look for altruists. In one pair, both problems begin with the following text:

> You are an anthropologist studying the Kaluame, a Polynesian people who live in small, warring bands on Maku Island in the Pacific. You are interested in how Kaluame "big men" —chieftains— wield power.
>
> "Big Kiku" is a Kaluame big man who is known for his ruthlessness. As a sign of loyalty, he makes his own "subjects" put a tattoo on their face. Members of other Kaluame bands never have facial tattoos. Big Kiku has made so many enemies in other Kaluame bands, that being caught in another village with a facial tattoo is, quite literally, the kiss of death.
>
> Four men from different bands stumble into Big Kiku's village starving and desperate. They have been kicked out of their respective villages for various misdeeds, and have come to Big Kiku because they need food badly. Big Kiku offers each of them the following deal:
>
> "If you get a tattoo on your face, then I'll give you cassava root."
>
> Cassava root is a very sustaining food which Big Kiku's people cultivate. The four men are very hungry, so they agree to Big Kiku's deal. Big Kiku says that the tattoos must be in place tonight, but that the cassava root will not be available until the following morning.

At this point the two problems diverge. The *cheater version* continues:

> You learn that Big Kiku hates some of these men for betraying him to his enemies. You suspect he will cheat and betray some of them. Thus, this is a perfect opportunity for you to see first hand how Big Kiku wields his power.
>
> The cards below have information about the fates of the four men. Each card represents one man. One side of a card tells whether or not the man went through with the facial tattoo that evening and the other side of the card tells whether or not Big Kiku gave that man cassava root the next day.
>
> Did Big Kiku get away with cheating any of these four men? Indicate only those card(s) you definitely need to turn over to see if Big Kiku has broken his word to any of these four men.

The *altruist version* continues:

> You learn that Big Kiku hates some of these men for betraying him to his enemies. You suspect he will cheat and betray some of them. However, you have also heard that Big Kiku sometimes, quite unexpectedly, shows great generosity towards others—that he is sometimes quite altruistic.

> Thus, this is a perfect opportunity for you to see first hand how Big Kiku wields his power.
>
> The cards below have information about the fates of the four men. Each card represents one man. One side of a card tells whether or not the man went through with the facial tattoo that evening and the other side of the card tells whether or not Big Kiku gave that man cassava root the next day.
>
> Did Big Kiku behave altruistically towards any of these four men? Indicate only those card(s) you definitely need to turn over to see if Big Kiku has behaved altruistically towards any of these four men.

The four cards, which were identical in both versions, were:

got the tattoo	Big Kiku gave him nothing
no tattoo	Big Kiku gave him cassava root

In the version of the problem that requires subjects to detect cheaters, Cosmides 1989 had found that 74% of subjects get the correct answer. In the version that requires subjects to detect altruists, however, only 28% answered correctly. (Cosmides and Tooby 1992, pp. 93-97).

These experiments, along with a number of others reviewed in Cosmides and Tooby 1992 are all compatible with the hypothesis that we have one or more Darwinian modules designed to deal with social exchanges and detect cheaters. However, this hypothesis is, to put it mildly, very controversial. Many authors have proposed alternative hypotheses to explain the data, and in some cases they have supported these hypotheses with additional experimental evidence. One of the most widely discussed of these alternatives is the *pragmatic reasoning schemas* approach defended by Cheng, Holyoak and their colleagues. (Cheng and Holyoak 1985 & 1989, Cheng, Holyoak, Nisbett and Oliver 1986). On this account, reasoning is explained by the activation of domain specific sets of rules (called "schemas") which are acquired during the lifetime of the individual through general inductive mechanisms. These rules subserve people's reasoning about permission, obligation, and other deontic concepts that may be used in their culture. Rules for reasoning about social exchanges are just one kind of reasoning schema. One virtue of this theory is that it provides an explanation for the fact that people perform well on problems like the bouncer-in-the-Boston-bar that are not comfortably assimilated to the model of reciprocal social exchange. However, as Cummins 1996 argues, there is little evidence for the claim that schemas involved in reasoning about permission and obligation are learned, and a fair amount of evidence suggesting that capacity to engage in deontic reasoning emerges relatively early in childhood. This, along with a number of other lines of evidence lead Cummins to propose an intriguing hypothesis that integrates ideas from both the social exchange theory and the pragmatic reasoning schemas theory. On Cummins' hypothesis, reasoning about "permissions, obligations, prohibitions, promises, threats and warnings" (p. 166) is subserved by an innate, domain specific module devoted exclusively to deontic contents. This reasoning module "evolved for the very important purpose of solving problems that frequently arise within a dominance hierarchy—the social structure that characterizes most mammalian and avian species."(p. 166.) A core component of the deontic reasoning module, Cummins maintains, is a mechanism whose job is violation detection. "[T]o reason effectively about deontic concepts, it is necessary to recognize what constitutes a violation, respond to it appropriately (which often depends on the respective status of the parties involved), and appreciate the necessity of adopting a violation-detection strategy whenever a deontic situation is encountered." (p. 166.) Still other hypotheses to account for the content effects in selection tasks have been proposed by Oaksford and Chater 1994, Manktelow and Over 1995 and Sperber, Cara and Girotto 1995.

This is not the place to review all of these theories, nor would we venture a judgment—even a tentative one—on which theory is most promising. These are busy and exciting times for those studying human reasoning, and there is obviously much that remains to be discovered. What we believe we can safely conclude from the studies recounted in this section is that the hypothesis that much of human reasoning is subserved by a cluster of domain specific Darwinian modules deserves to be taken very seriously. Whether or not it ultimately proves to be correct, the highly modular picture of the mechanisms underlying reasoning has generated a great deal of impressive research and will continue to do so for the foreseeable future. Thus we would do well to begin exploring what the implications would be for various claims about human rationality *if* the Massive Modularity Hypothesis turns out to be correct. In the final section of this paper we will begin this exploration by asking what implications the Massive Modularity Hypothesis might have for the "Bleak Implications" interpretation of some of the experimental studies of reasoning.

6. MASSIVE MODULARITY, BLEAK IMPLICATIONS AND THE PANGLOSSIAN INTERPRETATION

One possible response to the Massive Modularity Hypothesis—we'll call it the *Panglossian interpretation*—maintains that if MMH turns out to be correct, it would make the Bleak Implications interpretation of the experimental studies of rationality completely untenable. According to the Bleak Implications interpretation, the sorts of experimental results surveyed in Section 2 reflect shortcomings in human reasoning *competence*. People deal with the problems in those experiments by exploiting various normatively problematic heuristics, and they do this because they have nothing better available. They "lack the correct programs for many important judgmental tasks"[14] because, as Gould maintained, "our minds are not built ... to work by the rules of probability." (Gould 1992, p. 469.) But according to the Panglossian this is simply the wrong interpretation. If the Massive Modularity Hypothesis is correct, then the mind contains "a multitude of... elegant machines." (Cosmides & Tooby 1995, p. xiv.) There are Darwinian modules that reason in *normatively appropriate* ways about probability, cheating and threats and also about dead reckoning, intuitive mechanics, intuitive biology, intuitive psychology, and no doubt a host of others as well. So humans *do* have access to the correct programs for important judgmental tasks, our minds include Darwinian modules that *are* built to "work by the rules of probability," and humans are "good intuitive statisticians after all." The errors reported in the experimental literature, if indeed they really are errors,[15] are merely *performance* errors, and the Bleak Implications interpretation must be rejected.

We are not at all sure that anyone actually advocates this very strong version of the Panglossian interpretation, though we suspect that a fair number of people would endorse a more hedged and cautious version.[16] We don't believe that anything very close to the strong version of the Panglossian interpretation can be defended, though we think there is a great deal to be learned by exploring why the Panglossian interpretation fails.

One fairly straightforward objection to the Panglossian interpretation begins with the observation that the experimental literature on human reasoning has documented many quite different sorts of problems on which subjects perform poorly. Those reviewed in Section 2 are a small and highly selective sample. If the Panglossian interpretation is correct, then people must have Darwinian modules capable of handling in normatively appropriate ways *all* of the problems on which subjects perform poorly, though for one reason or another the performance of experimental subjects does not reflect their underlying competence. That is, of course, a very strong claim, much stronger than currently available evidence will support. Nor is there any plausible evolutionary argument for the claim that natural selection would have provided us with Darwinian modules for handling *all* of these cases. So the Panglossian interpretation rests on a bold speculation with relatively little empirical or theoretical support. But even if we put this concern off to the side and concentrate on those cases where there is some evidence for the existence of a Darwinian module, there are serious problems with the Panglossian idea that all errors are performance errors.

To bring these problems into focus, let us start by considering Kahneman and Tversky's seven-letter-word problem, discussed in Sec. 5.1. In that problem subjects were not asked about the probability of a particular event. Rather, they were asked to estimate the *frequency* of words for the form '----ing' and words of the form '-----n-' in four pages of text. Yet despite being asked to estimate frequencies, most subjects said that the number of '----ing' words would be greater than the number of '-----n-' words. If, as advocates of MMH have argued, we have one or more Darwinian modules that do a good job of probabilistic reasoning when problems are couched in terms of frequencies, what sort of explanation can be offered for the error that these subjects make? One plausible hypothesis is that rather than using their probabilistic reasoning module(s), subjects are relying on what Kahneman and Tversky call an "availability heuristic." They are searching memory for examples of words of the form '----ing' and also for words of the form '-----n-' and, because of the way in which our memory for such facts is organized, they are coming up with far more of the former than of the latter. But now let us ask *why* subjects (or their cognitive systems) are dealing with the problem in this way. Why *aren't* they using a probabilistic reasoning module which, presumably, would not produce responses that violate the conjunction rule? For an advocate of MMH, perhaps the most natural hypothesis is that there is a mechanism in the mind (or maybe more than one) whose job is to determine which of the many reasoning modules and heuristics that are available in a Massive Modular mind get called on to deal with a given problem, and that this mechanism, which we'll call *the allocation mechanism*, is routing the problem to the wrong component of the reasoning system. If that's right, and if we further suppose that this mis-allocation is the result of persisting and systematic features of the allocation mechanism, then it seems natural to conclude that the allocation mechanism itself is normatively problematic. It produces errors in reasoning by sending problems to the wrong place.

If this speculation is correct—if certain errors in reasoning are generated by a normatively problematic allocation mechanism—then it seems odd to say that the resulting errors are "performance errors." For unlike performance errors that result from fatigue or alcohol or emotional stress, this is not a case in which factors arising outside the reasoning system interfere with the normal functioning of the system and cause it to operate in a way that it does not usually operate. In dealing with cases like the seven-letter-word problem, the allocation mechanism works just the way it normally does. The reasoning error is produced because what it normally does is send problems like these to the wrong place. Nor does this look much like the sorts of performance errors that are produced in language processing as the result of limited short term memory. There is no resource that runs out in these cases of mis-allocation, no parameter that is exceeded. The subject gets the wrong answer because the principles governing the operation of the allocation system are themselves normatively defective. There is (we have been assuming) a Darwinian module capable of doing a good job on the problem, and the allocation mechanism fails to send it there. At this point, a defender of the Panglossian interpretation might insist that since the correct rules for handling these cases of faulty reasoning are available in the subject's mind, the errors are not the product of a defective competence, and thus allocation errors must be just *another kind* of performance error. This argument assumes that there are only two kinds of cognitive errors—performance errors and competence errors—and that anything which doesn't count as one sort of error must be an instance of the other sort. But that is not an assumption we see any reason to accept. Since mis-allocation errors are not comfortably viewed either as competence errors or as performance errors, we are inclined to think that one lesson to be learned from examples like this is that in a Massively Modular Mind the performance error/competence error distinction does not exhaust the possibilities.

Let us turn, now, to the original version of the feminist bank teller problem (Sec. 2.2) and the original version of the Casscells et al. "Harvard Medical School" problem (Sec. 2.3). In both cases subjects perform poorly. How might an advocate of the Massive Modularity Hypothesis explain this poor performance? One possibility is that these are further examples of allocation errors, and that there is a reasoning module that would have solved them correctly had they been routed there. But the is also a very different possibility that needs to be explored. Darwinian modules are designed by natural selection to handle recurrent information processing problems. To enable a module to handle problems efficiently, one strategy that natural selection might exploit is to design the module in such a way that it can deal successfully with a problem only if the problem is presented in an appropriate format or in an appropriate system of representation. Thus, for example, Gigerenzer argues that since frequentist formats were the only ones to play a major role in the EEA, we would expect the mental module(s) that handle probabilistic reasoning be designed to "expect" that format and to be unable to solve the problems successfully if they are presented in some other format. If Gigerenzer is right, then the module(s) subserving good bayesian reasoning simply cannot solve problems posed in terms of single event probabilities. But in that case, subjects' errors in the original version of the Harvard Medical School problem and the feminist bank teller problem cannot be treated as allocation errors, since the allocation system hasn't sent them to the wrong place. It has no good place to send them. In ordinary subjects there is no module or component of the reasoning system that has the right algorithms for dealing with the problem as posed.

If these speculations are right, then it might be tempting to conclude that the errors are competence errors, and thus that the Bleak Implications interpretation has gained a foothold even within a Massively Modular picture of the mind. But, while the matter may be largely terminological, we are not entirely comfortable with the conclusion that these errors are competence errors. For while it is true that the hypothesized Darwinian module(s) don't contain algorithms that can deal with the problems *as posed*, it is also the case that the modules do contain algorithms for dealing with *reformulated* versions of the problems. Thus it may be possible to improve people's performance on these problems without modifying their competence and enriching the reasoning algorithms that the mind makes available. For we may be able to teach them to restate the problems, putting them into a format that their Darwinian modules are designed to process. Since the distinction between those errors that can be avoided by reformulation and those that cannot is potentially a very important one, we think the avoidable errors merit a category of their own. We'll call them *formulation errors.*

One central claim made by the Panglossian interpretation is that all the errors reported in the experimental literature are merely performance errors. But we've now seen two quite different reasons to be suspicious of that claim. If the Massive Modularity Hypothesis is correct then some reasoning errors are likely to be mis-allocation errors, while others may be formulation errors. On our view, the right conclusion to draw from the Massive Modularity Hypothesis is not that all errors are performance errors, but rather that there are a number of importantly different kinds of errors that can't be comfortably characterized as either performance errors or competence errors. If MMH is right, then the assumption that all reasoning errors are either performance errors or competence errors will have to be abandoned.

The other central claim made by the Panglossian interpretation is that the mind is well stocked with Darwinian modules that reason in normatively appropriate ways. In the remaining pages of this chapter we want to consider some of the problems that confront this component of the Panglossian interpretation. A first problem is settling on what might be called a *general normative theory of reasoning*—a theory which specifies the standards by which any inference mechanism or reasoning strategy should be evaluated. In the philosophical literature there is a great deal of debate about the attractions of competing general normative theories.[17] Some theorists defend "reliabilist" accounts in which attaining true beliefs plays a central role. Others advocate accounts on which attaining more pragmatic goals like health and happiness are central. Still others urge that reasoning strategies should be evaluated by appeal to our reflective intuitions about what is and is not rational. This is not the place to review the arguments for and against these general normative theories. Rather, we will assume, as we have throughout this chapter, that some version of reliabilism is correct and that truth is central to the evaluation of inferential mechanisms. Other things being equal, one inferential mechanism is better than another if it does a better job at getting the right answer. But even if we assume that reliabilism is the correct general normative theory of reasoning, the domain specificity of Darwinian modules poses a cluster of new and quite unique problems that traditional epistemology has not yet explored.

Consider, for example, the module that subserves reasoning about social contracts. We can assume that this module does a relatively good job at answering questions about cheating and contract violation. But there are also indefinitely many problems—elementary arithmetic problems, for example, or "theory of mind" problems about what people would believe or decide to do in various circumstances—for which the social contract module does not produce the right answer; indeed, it produces no answer at all. But surely it would be perverse to criticize the social contract module on the grounds that it can't solve mathematical problems. This would be a bit like criticizing a toaster on the grounds that it cannot be used as a typewriter. To evaluate a toaster we must attend to its performance on an appropriate range of tasks, and clearly typing is not one of them. Similarly, to evaluate the social contract module we must attend to its performance on an appropriate range of tasks, and solving math problems is not one of them. The moral to be drawn here seems fairly obvious: Normative evaluations of domain specific modules must be relativized to a specific domain or a specific range of problems. But this immediately raises a new puzzle: If normative evaluations of domain specific modules must be relativized to a domain, which domain should it be?

One suggestion is that the right domain is what Sperber 1994 calls the *actual domain*. The actual domain for a given reasoning module is "all the information in the organism's environment that (once processed by perceptual modules, and possibly by other conceptual modules) satisfy the module's input conditions." (p. 52.) By "input conditions" Sperber means those conditions that must be satisfied in order that the module be able to process a given item of information. So, for example, if a module requires that a problem be stated in a particular format, then any information not stated in that format fails to satisfy the module's input conditions.

A quite different suggestion is that the domain relevant to the evaluation of domain specific modules is what Sperber calls the *proper domain*, which he characterizes as "all the information that it is the module's biological function to process." (p. 52.) The proper domain is the information that the module was designed to process by natural selection. In recent years, many philosophers of biology have come to regard the notion of a biological function as a particularly slippery one.[18] For current purposes we can rely on the following very rough characterization: The biological functions of a system are the activities or effects of the system in virtue of which it has remained a stable feature of an enduring species.

In some cases the actual domain of a Darwinian module may coincide with its proper domain. But it is also likely that in many cases the two domains will not be identical. For example, it is plausible to suppose that the proper domain of the folk psychology module includes only the kind of information about the mental states of human beings, and the behavior caused by those states, that would have been useful to our Pleistocene forebears. But it is very likely that the module also processes information about lots of other things including the activities of non-human animals, cartoon characters and even mindless physical objects like trees and heavenly bodies. If this is right, then a normative evaluation of the module relativized to its proper domain is likely to be much more favorable than a normative evaluation relativized to its actual domain. We suspect that those Panglossian inclined theorists who describe Darwinian modules as "elegant machines" are tacitly assuming that normative evaluation should be relativized to the proper domain, while those who offer a bleaker assessment of human rationality are tacitly relativizing their evaluations to the actual domain which, in the modern world, contains a vast array of information processing challenges that are quite different from anything that our Pleistocene ancestors had to confront.

So which domain should we use in to evaluate the module, the proper domain or the actual one? Which domain is the *right* one? We don't think there is any principled way of answering this question. Rather, we maintain, normative claims about Darwinian modules or the algorithms they embody, make no clear sense until they are explicitly or implicitly relativized to a domain. Moreover, the choice confronting us is actually much more complex than we have so far suggested. For both actual domains and proper domains are best viewed not as single options but as families of options. There are different ways of explicating both the notion of a proper domain and the notion of an actual domain, and these differences will make a difference, in some cases a major difference, in the outcome of relativized normative assessments. (See Samuels, in preparation.) Nor should it be assumed that actual domains and proper domains are the only two families of options that might be considered. Normative assessments can serve many different purposes, and for some of these it may be appropriate to relativize to a domain which is neither actual nor proper.

Our conclusion is that neither the Panglossian interpretation nor the Bleak Implications interpretation offers a satisfactory response to the Massive Modularity Hypothesis. If it is indeed the case that our minds contain a large number of Darwinian modules, and that the modules subserve most of our everyday reasoning, then many of the categories and distinctions that philosophers and cognitive scientists have used to describe and assess cognition will have to be reworked or abandoned. If the Massive Modularity Hypothesis is correct, we will have to rethink what we mean by "rationality."

R. Samuels, S. Stich, P. D. Tremoulet
Rutgers University
U.S.A..

NOTES

* Earlier versions of some of this material served as the basis of lectures at The City University of New York Graduate Center, Canterbury University in Christchurch New Zealand, Rutgers University and at the 5th International Colloquium on Cognitive

Science in San Sebastian, Spain. We are grateful for the many helpful comments and criticisms that were offered on these occasions. Special thanks are due to Kent Bach, Michael Bishop, Margaret Boden, Derek Browne, L. Jonathan Cohen, Jack Copeland, Stephen Downes, Mary France Egan, Richard Foley, Gerd Gigerenzer, Daniel Kahneman, Ernie LePore, Brian McLaughlin, Brian Scholl, and Ernest Sosa.

[1] For references see L.J. Cohen, *The Dialogue of Reason*, Oxford: Clarendon Press, 1986, p. 150ff.

[2] R.E. Nesbitt and E. Borgida, "Attribution and the psychology of prediction", *Journal of Personal and Social Psychology* 32, 1975, 932-943.

[3] See L.J. Cohen, "Can human irrationality be experimentally demonstrated?", *The Behavioral and Brain Sciences* 4, 1981, 317-370, and G. Gigerenzer, "How to make cognitive illusions disappear: beyond 'Heuristica and Biases'", in W. Stroebe and Miles Hewstone (eds.), *European Review of Social Psychology*, vol. 2, 1991, 83-115.

[4] A. Tversky, and D. Kahneman, "Judgement under uncertainty: heuristics and biases", *Science* 125, 1974, 1124-1131.

[5] A. Tversky, and D. Kahneman, "On the psychology of prediction", *Oregon Research Institute Research Bulletin* 12, 1972, 4.

[6] A. Tversky, and D. Kahneman, "Judgement under uncertainty: heuristics and biases", *Science* 125, 1974, 1124-1131.

[7] C.J. Jones, and P.L. Harris, "Insight into the law of large numbers: a comparison of Piagetian and judgement theory", *Quarterly Journal of Experimental Psychology*, 34A, 1982, 479-488.

[8] See L.J. Cohen, "Twelve questions about Keynes's concept of weight", *British Journal for the Philosophy of Science* 37, 1986, 263-278.

[9] See, for example, L.J. Cohen, *The Probable and the Provable*, Oxford: Clarendon Press, 1970.

[10] M. Bar-Hillel, "On the subjective probability of compound events", in *Organisational Behaviour and Human Performance* 9, 1973, 396-406.

[11] I. Hacking, *The Emergence of Probability: a Philosophical Study of Early Ideas about Probability, Induction and Statistical Inference*, Cambridge: Cambridge University Press.

[12] Transactions in Human Factors in Electronics HFE-7, 1966, 29-37; and L.R. Beach and C.R. Peterson, 'Subjective probabilities for unions of events', *Psychonomic Science* 5, 1966, 307-308.

[13] L.J. Cohen, "Probability—the one and the many", *Proceedings of the British Academy* LXI, 1975, 83-108.

[14] H.P. Grice, "Logic and conversation", in D. Davidson and G. Harman (eds.), *The Logic of Grammar*, Encino, California: Dickinson, 1975, 64-75.

[15] S. Scribner, "Modes of thinking and ways of speaking: culture and logic reconsidered", in P.N. Johnson-Laird and P.C. Wason (eds.), *Thinking; Readings in Cognitive Science*, Cambridge: Cambridge University Press, 1977, 483-500.

[16] P.C. Wason, "Reasoning", in B. Foss (ed.), *Psychology*, Harmondsworth: Penguin, New Horizons, 1966, 135-51.

[17] P.C. Wason, and D. Shapiro, "Natural and contrived experience in a reasoning problem", *Quarterly Journal of Experimental Psychology* 23, 1979, 63-71

[18] Notably R. Carnap, *Logical Foundations of Probability*, Chicago: Chicago University Press, 1950, 19-51. See also the references given in L.J. Cohen, *An Introduction to the Philosophy of Induction and Probability*, Oxford: Clarendon Press, 1989, 42.

REFERENCES

Barkow, J. 1992: Beneath new culture is old psychology: Gossip and social stratification. In Barkow, Cosmides and Tooby 1992, 627-637.

Barkow, J., Cosmides, L., and Tooby, J. (eds.), 1992: *The Adapted Mind: Evolutionary Psychology and the Generation of Culture*. Oxford: Oxford University Press.

Baron, J. 1988: *Thinking and Deciding*. Cambridge: Cambridge University Press.

Baron-Cohen, S. 1994: How to build a baby that can read minds: Cognitive mechanisms in mindreading. *Cahiers de Psychologie*, ***13***, 5, 513-552.

Baron-Cohen, S. 1995: *Mindblindness: An Essay on Autism and Theory of Mind*. Cambridge, MA: MIT Press.

Carey, S. and Spelke, E. 1994: Domain-specific knowledge and conceptual change. In Hirschfeld and Gelman 1994, 169-200.

Carruthers, P. and Smith, P. (eds.), 1996: *Theories of Theories of Mind*. Cambridge: Cambridge University Press.

Casscells, W., Schoenberger, A. and Grayboys, T. 1978: Interpretation by physicians of clinical laboratory results. *New England Journal of Medicine*, ***299***, 999-1000.

Chapman, L. and Chapman, J. 1967: Genesis of popular but erroneous diagnostic observations. *Journal of Abnormal Psychology*, ***72***, 193-204.

Chapman, L. and Chapman, J. 1969: Illusory correlation as an obstacle to the use of valid psychodiagnostic signs. *Journal of Abnormal Psychology*, ***74***, 271-280.

Cheng, P. and Holyoak, K. 1985: Pragmatic reasoning schemas. *Cognitive Psychology*, *7*, 391-416.

Cheng, P. and Holyoak, K. 1989: On the natural selection of reasoning theories. *Cognition*, ***33***, 285-313.

Cheng, P., Holyoak, K., Nisbett, R., and Oliver, L. 1986: Pragmatic versus syntactic approaches to training deductive reasoning. *Cognitive Psychology*, ***18***, 293-328.

Chomsky, N. 1965: *Aspects of the Theory of Syntax*. Cambridge, MA: MIT Press.

Chomsky, N. 1975: *Reflections of Language*. New York: Pantheon Books.

Chomsky, N. 1980: *Rules and Representations*. New York: Columbia University Press.

Cohen, L. 1981: Can human irrationality be experimentally demonstrated? *Behavioral and Brain Sciences*, ***4***, 317-370.

Cohen, L. 1986: *The Dialogue of Reason*. Oxford: Clarendon Press.

Cosmides, L. 1989: The logic of social exchange: Has natural selection shaped how humans reason? Studies with Wason Selection Task. *Cognition*, ***31***, 187-276.

Cosmides, L. and Tooby, J. 1992: Cognitive adaptations for social exchange. In Barkow, Cosmides and Tooby 1992, 163-228.

Cosmides, L. and Tooby, J. 1994: Origins of domain specificity: The evolution of functional organization. In Hirschfeld and Gelman 1994, 85-116.

Cosmides, L. and Tooby, J. 1996: Are humans good intuitive statisticians after all? Rethinking some conclusions from the literature on judgment under uncertainty." *Cognition*, ***58***, 1, 1-73.

Cummins, D. 1996: Evidence for the innateness of deontic reasoning. *Mind and Language*, ***11***, 160-190.

Dawes, R. 1988: *Rational Choice in an Uncertain World*. Orlando, FL: Harcourt Brace Jovanovich.

Dawes, R. 1994: *House of Cards: Psychology and Psychotherapy Built on Myth*. New York: Free Press.

Fiedler, K. 1988: The dependence ofthe conjunction fallacy on subtle linguistic factors. *Psychological Research*, ***50***, 123-129.

Fodor, J. 1983: *The Modularity of Mind*. Cambridge, MA: MIT Press.

Fodor, J. 1986: The modularity of mind. In Pylyshyn and Demopoulos 1986, 3-18.
Gallistel, C. 1990: *The Organization of Learning*. Cambridge, MA: MIT Press.
Gardner, H. 1983: *Frames of Mind: The Theory of Multiple Intelligences*. New York: Basic Books.
Garfield, J. (ed.), 1987: *Modularity in Knowledge Representation and Natural-Language Understanding*. Cambridge, MA: MIT Press.
Gelman, S. and Brenneman K. 1994: First principles can support both universal and culture-specific learning about number and music. In Hirschfeld and Gelman 1994, 369-387):
Gigerenzer, G. 1991: How to make cognitive illusions disappear: Beyond 'heuristics and biases.'*European Review of Social Psychology*, *2*, 83-115.
Gigerenzer, G. 1994: Why the distinction between single-event probabilities and frequencies is important for psychology (and vice versa): In G. Wright and P. Ayton, eds., *Subjective Probability*. New York: John Wiley.
Gigerenzer, G. and Hug, K. 1992: Domain-specific reasoning: Social contracts, cheating and perspective change. *Cognition*, ***43***, 127-171.
Gigerenzer, G., and Hoffrage, U. 1995: How to improve Bayesian reasoning without instruction: Frequency formats. *Psychological Review*, ***102***, 684-704.
Gigerenzer, G., Hoffrage, U., and Kleinbölting, H. 1991: Probabilistic mental models: A Brunswikean theory of confidence. *Psychological Review*, ***98***, 506-528.
Gigerenzer, G. and Murray, D. 1987: *Cognition as Intuitive Statistics*, Hillsdale, NJ: Erlbaum.
Gilovich, T., Vallone, B. and Tversky, A. 1985: The hot hand in basketball: On the misconception of random sequences. *Cognitive Psychology*, ***17***, 295-314.
Gluck, M. and Bower, G. 1988: From conditioning to category learning: An adaptive network model. *Journal of Experimental Psychology: General*, ***117***, 227-247.
Godfrey-Smith, P. 1994: A modern history theory of functions. *Nous*, ***28***, 344-362.
Goldman, A. 1986: *Epistemology and Cognition*, Cambridge, MA: Harvard University Press.
Gould, S. 1992: *Bully for Brontosaurus. Further Reflections in Natural History*. London: Penguin Books.
Griggs, R. and Cox, J. 1982: The elusive thematic-materials effect in Wason's selection task. *British Journal of Psychology*, ***73***, 407-420.
Griffiths, P. 1997: *What Emotions Really Are*. Chicago: The University of Chicago Press.
Haugeland, J. 1985: *Artificial Intelligence: The Very Idea*. Cambridge, MA: MIT Press.
Hertwig, R. and Gigerenzer, G. 1994: The chain of reasoning in the conjunction task. Unpublished manuscript.
Hirschfeld, L. and Gelman, S. (eds.), 1994: *Mapping the Mind*. Cambridge: Cambridge University Press.
Hutchins, E. 1980: *Culture and Inference: A Trobriand Case Study*. Cambridge, MA: Harvard University Press.
Jackendoff, R. 1992: Is there a faculty of social cognition? In R. Jackendoff, *Languages of the Mind*. Cambridge, MA: MIT Press, 69-81.
Kahneman, D. and Tversky, A. 1973: On the psychology of prediction. *Psychological Review*, ***80***, 237-251. Reprinted in Kahneman, Slovic and Tversky 1982:
Kahneman, D. and Tversky, A. 1996: On the reality of cognitive illusions. *Psychological Review*, ***103***, 582-591.
Kahneman, D., Slovic, P. and Tversky, A. 1982: *Judgment Under Uncertainty: Heuristics and Biases*. Cambridge: Cambridge University Press.
Karmiloff-Smith, A. 1992: *Beyond Modularity: A Developmental Perspective on Cognitive Science*. Cambridge, MA: MIT Press.

Lehman, D, Lempert, R and Nisbett, R. 1988: The effects of graduate education on reasoning: Formal discipline and thinking about everyday life events. *American Psychologist*, ***43***, 431-443.

Lehman, D. and Nisbett, R. 1990: A longitudinal study of the effects of undergraduate education on reasoning. *Developmental Psychology*, ***26***, 952-960.

Leslie, A. 1994: ToMM, ToBY, and agency: Core architecture and domain specificity. In Hirschfeld and Gelman 1994, 119-148.

Lichtenstein, S., Fischoff, B. and Phillips, L. 1982: "Calibration of probabilities: The state of the art to 1980. In Kahneman, Slovic and Tversky 1982, 306-334.

Manktelow, K. and Over, D. 1995: Deontic reasoning. In S. Newstead and J. St. B. Evans, eds, *Perspectives on Thinking and reasoning*. Hillsdale, N.J.: Erlbaum.

Neander, K. 1991: The teleological notion of 'function'. *Australasian Journal of Philosophy*, ***59***, 454-468.

Nisbett, R. and Ross, L. 1980: *Human Inference: Strategies and Shortcomings of Social Judgment*. Englewood Cliffs, NJ: Prentice-Hall.

Nisbett, R., Fong, G, Lehman, D. and Cheng, P. 1987: Teaching reasoning. *Science*, ***238***, 625-631.

Oaksford, M. and Chater, N. 1994: A rational analysis of the selection task as optimal data selection.*Psychological Review*, ***101***, 608-631.

Piattelli-Palmarini, M. 1994: *Inevitable Illusions: How Mistakes of Reason Rule Our Minds*. New York: John Wiley & Sons.

Pinker, S. 1994: *The Language Instinct*. New York: William Morrow and Co.

Pinker, S. 1997: *How the Mind Works*. New York: W. W. Norton.

Plantinga, A. 1993: *Warrant and Proper Function*. Oxford: Oxford University Press.

Pylyshyn, Z. 1984: *Computation and Cognition*. Cambridge, MA: MIT Press.

Pylyshyn, Z. and Demopoulos, W. (eds.), 1986: *Meaning and Cognitive Structure: Issues in the Computational Theory of Mind*. New Jersey: Ablex.

Quartz, S. and Sejnowski, T. 1994: Beyond modularity: Neural constructivist principles in evelopment. *Behavioral and Brain Sciences*, ***17***, 725-726.

Samuels, R. (in preparation a): Evolutionary psychology and the massive modularity hypothesis.

Samuels, R. (in preparation b): How to dissolve the rationality debate.

Segal, G. 1996: The modularity of theory of mind. In Carruthers and Smith 1995, 141-157.

Slovic, P., Fischhoff, B., and Lichtenstein, S. 1976: Cognitive processes and societal risk taking. In J. S. Carol and J. W. Payne, eds, *Cognition and Social Behavior*. Hillsdale, NJ: Erlbaum.

Sperber, D. 1994: The modularity of thought and the epidemiology of representations. In Hirschfeld and Gelman 1994, 39-67.

Sperber, D., Cara, F. and Girotto, V. 1995: Relevance theory explains the selection task. *Cognition*, ***57***, 1, 1995.

Stein, E. 1996: *Without Good Reason*. Oxford: Clarendon Press.

Stich, S. 1990: *The Fragmentation of Reason*. Cambridge, MA: MIT Press.

Sutherland, S. 1994: *Irrationality: Why We Don't Think Straight!* New Brunswick, NJ: Rutgers University Press.

Tanenhaus, M., Dell, G., and Carlson, G. 1987: Context effects and lexical processing: A connectionist approach to modularity. In Garfield 1987, 83-108.

Tooby, J. and Cosmides, L. 1992: The psychological foundations of culture. In Barkow, Cosmides and Tooby 1992, 19-136.

Tooby, J. and Cosmides, L. 1995: Foreword. In Baron-Cohen 1995, xi-xviii.

Trivers, R. 1971: The evolution of reciprocal altruism. *Quarterly Review of Biology*, ***46***, 35-56.

Tversky, A. and Kahneman, D. 1983: Extensional versus intuitive reasoning: The conjunction fallacy in probability judgement. *Psychological Review*, ***90***, 293-315.

Tversky, A. and Kahneman, D. 1982: Judgments of and by representativeness. In Kahneman, Slovic and Tversky 1982, 84-98.

Wilson, M. and Daly, M. 1992: The man who mistook his wife for a chattel. In Barkow, Cosmides and Tooby 1992, 289-322.

RUTH M.J. BYRNE, ALESSANDRA TASSO, VALERIE THOMPSON

THE PSYCHOLOGY OF COUNTERFACTUAL CONDITIONALS

1. COUNTERFACTUAL THINKING

In our daily thinking we often make inferences that are based not only on the facts of a situation but also on conjectures about how the situation could have been different. Everyday reasoning may be about current facts, e.g.:

Seamus Heaney receives the Nobel prize for literature.

or past facts, e.g.:

Samuel Beckett received the Nobel prize for literature.

Everyday thinking about nonfactual or hypothetical states of affairs may be based on present possibilities (that could happen given the actual state of the world), e.g.:

Roddy Doyle receives the Nobel prize for literature.

or on past possibilities (that could have happened given the actual state but did not), e.g.:

James Joyce received the Nobel prize for literature.

(see Johnson-Laird & Byrne, 1991, Chapter 4; Byrne, 1997.) Everyday thinking about nonfactual states of affairs may also be based on impossibilities (that could never happen in the past or present given the actual state of the world), such as:

Seamus Heaney receives the Nobel prize for science.

K. Korta et al. (eds.), Cognition, Agency and Rationality, 63-77.

States of affairs that deal with past possibilities and present or past impossibilities are known as counterfactual states of affairs (see also Byrne, 1997; Johnson-Laird & Byrne, 1991).

Thinking about what might be or what might have been is unique in that it requires reasoners to represent what is false, temporarily supposed to be true. People often engage in counterfactual thinking about what might have been in their everyday thoughts. If your car is stolen while parked in a rough area of the city, you might think about how things could have been different, e.g.,

If I had parked elsewhere my car wouldn't have been stolen.

When we think about what might have been we often mentally undo a bad outcome, and then mentally undo the perceived causes of that outcome to create a scenario in which the bad outcome did not occur (e.g., Kahneman & Miller, 1986). Counterfactual thinking may help us to learn from our mistakes by helping us identify what led to them, and to formulate plans about how to avoid their recurrence (e.g., Roese, 1994). In addition, it plays a role in other sorts of higher-order cognition, for example, in reasoning to form counterexamples to conclusions (e.g., Johnson-Laird & Byrne, 1991), in problem-solving to form sub-goals (e.g., Ginsberg, 1991; Keane, 1997), and in creativity to formulate the intention to improve (e.g., Hofstadter, 1979). Counterfactual thinking can also be an interface between cognition and emotion, and emotions such as regret, guilt, or relief may arise because of the comparison between the way a situation actually turned out, and the way it might have been different (e.g., Landman, 1987; Gilovich & Medvec, 1994).

How do people think about what might have been? To answer the question we need to examine the cognitive mechanisms that underlie reasoning about matters of fact and reasoning about matters of possibility and impossibility. There are currently several alternative theories of reasoning, developed to account for reasoning about matters of fact. One theory is that people rely on formal inference rules (e.g., Braine & O'Brien, 1991; Rips, 1994). According to this theory, the internal symbols in the mind, or mental representations, that reasoners rely on are language-like, and the operations that manipulate these symbols, or cognitive processes, are based on syntactic rules, and proofs of conclusions. A second theory is that people rely on domain-specific inference rules (e.g., Cheng & Holyoak, 1985; Cosmides, 1989). According to this view, the mental representations that reasoners rely on are, once again, language-like, and the cognitive processes are based on content-sensitive rules. A third alternative is that people rely on mental models (e.g., Johnson-Laird, 1983; Johnson-Laird & Byrne, 1991). According to this view, the mental representations that reasoners rely on are models, that is, representations that correspond to the structure of the world not to the structure of the language used to describe the world, and the cognitive processes are based on rules that manipulate models. Although there are several competing theories to account for reasoning about matters of fact, it is only this last theory that has been extended to deal with reasoning about matters of possibility and impossibility and experimentally tested in these domains (Byrne & Tasso, in press; Johnson-Laird & Byrne, 1991; Johnson-Laird, Legrenzi, Girotto, Legrenzi, & Caverni, 1998). We will focus on the mental model theory in this chapter, and we will propose that it is possible for a theory of reasoning to encompass reasoning about matters of fact and reasoning about matters of possibility and impossibility (e.g., Johnson-Laird & Byrne, 1991).

We will outline our arguments with respect to conditionals, such as:

If Vittorio was in Trieste then Paolo was in Milan.

We will first sketch the challenge posed to a general theory of conditionals by counterfactual conditionals, and we will then outline a mental model theory of factual and counterfactual conditionals. We will describe experiments that test the predictions of this theory about the inferences that people make.

1.1. The Challenge of Counterfactual Conditionals

A factual conditional such as:

If Vittorio was in Trieste then Paolo was in Milan.

is in the indicative mood (and, in this example, in the past tense) whereas a counterfactual conditional is in the subjunctive mood (and, again, in the past tense):

If Vittorio had been in Trieste then Paolo would have been in Milan.

Philosophers have long recognized that counterfactual conditionals pose a challenge to a theory of conditionals to encompass them (e.g., Ramsey, 1931; Stalnaker, 1968; Lewis, 1973; Pollock, 1986). The counterfactual conditional appears to mean something very different from the factual conditional. The counterfactual conditional seems to presuppose that its antecedent is false, Vittorio is NOT in Trieste, and its consequent is false, Paolo is NOT in Milan either. The factual conditional makes no such presupposition.

The problem of counterfactual conditionals can be seen clearly if one considers in what situations a counterfactual could be false. For a factual conditional, even a truth-functional semantics offers a simple answer: the factual conditional is false in the following situation:

Vittorio is in Trieste and Paolo is not in Milan.

Of course there are cases where even factual conditionals cannot be accounted for by this simple truth-functional account (e.g., Johnson-Laird, 1986) and psychologists have long studied comprehension and deduction from factual conditionals (e.g., Byrne, 1989a; 1989b; Thompson, 1994; 1995; Girotto, Mazzocco & Tasso, 1997). But for many factual conditionals there is at least one clear case where they can be judged to be false: a conditional is false when its antecedent is true and its consequent is false. In contrast, there is no such clear case for a counterfactual conditional. The counterfactual presupposes that its antecedent is false from the outset, and so the state of affairs where its antecedent is true and its consequent is false is ruled out. Is there any other state of affairs in which a counterfactual conditional could be considered false? There seems to be no simple answer on a truth functional account. Of course, counterfactual conditionals are not all true, and people can readily distinguish between counterfactual conditionals that are plausible and those that are not (e.g., Miyamoto and Dibble, 1986).

The distinction between factual and counterfactual conditionals also applies in fictional discourse. Within the fictional world of Sherlock Holmes, the following counterfactual seems likely to be true:

If Sherlock Holmes had married, Dr. Watson would have been jealous.[1]

Counterfactual conditionals allow us to consider alternatives to what we are *currently treating* as the facts. A conditional can combine factual and counterfactual matters, in the antecedent or consequent: it may contain a factual antecedent and consequent, or it may contain a counterfactual antecedent and consequent, as in the examples considered earlier. A conditional may also combine a counterfactual antecedent with a factual consequent, e.g.,

If you had needed some money, there was some in the bank.

(Johnson-Laird, 1986), or a factual antecedent with a counterfactual consequent, e.g.,

If you needed money, I would have put some in the bank for you.

The antecedent and consequent can combine past, present, or future scenarios (e.g., Lewis, 1979). It is possible to refer to future possibilities, e.g.:

If Yeltsin were healthy next year then the Party would run smoothly.

and to situations that are not possible given the world as it is, e.g.:

If Stalin were still alive then the Party would run smoothly.

People clearly believe that counterfactuals convey the presupposition that their antecedents and consequents are false, and this belief has been demonstrated by one ofthe first psychologists to examine counterfactual conditionals (Fillenbaum, 1974). Fillenbaum examined counterfactual conditionals such as:

If he had caught the plane he would have arrived on time.

He suggested that the negation ofthe antecedent is presupposed, so much so that it appears inconsistent to assert the following:

If he had caught the plane, which he did, he would have arrived on time.

He also suggested that the negation of the consequent is strongly suggested but not required, so that there is nothing in principle contradictory about asserting:

If he had caught the plane he would have arrived on time, which he did anyway because he took a fast train.

Fillenbaum asked subjects to read conditionals and simple categorical propositions, and then he gave them an unexpected recognition test: they had to judge from a new set of conditionals and categorical propositions, which propositions they had seen in the previous set, and which ones were new. He found that the participants mistakenly 'recognized' on 25% of occasions categorical propositions of the following sort:

He did not catch the plane.

which corresponded to the negated antecedent of the counterfactual conditional they had seen before. Even more strikingly, the participants mistakenly recognized 44% of the negated consequents, such as:

He did not arrive on time.

even though the participants had never been given these propositions in the earlier set either. Counterfactual conditionals clearly convey the presupposition that their antecedents and consequents are false. How can such conditionals be encompassed within a general theory of conditionals? We will turn now to the model theory of conditionals to provide one answer to this question.

2. THE MODEL THEORY OF CONDITIONALS

The model theory of conditionals (Johnson-Laird & Byrne, 1991) proposes that a factual conditional, such as:

If Vittorio was in Trieste then Paolo was in Milan.

is understood by constructing a set of models which represent the way the world would be if the assertion was true:

Vittorio	Paolo
not-Vittorio	not-Paolo
not-Vittorio	Paolo

where "Vittorio" represents Vittorio in Trieste, "Paolo" represents Paolo in Milan, and "not" is a propositional-like tag to represent negation, so "not-Vittorio" represents Vittorio not in Trieste. Separate models are represented on separate lines and the information in the set of models for the conditional is represented in a fully explicit way. Models represent what is true, so the situation that is ruled out by the conditional, where Vittorio is in Trieste and Paolo is not in Milan, is not represented in the models. The models may include information about who Vittorio and Paolo are, what Trieste and Milan are like, why they may go to these places, and so on; the content of the models is not our concern here, their structure is.

According to the model theory, people may find it difficult to keep multiple models in mind because of the constraints of working memory and so they represent as much information as possible implicitly. The initial set of models may be as follows:

V	P
. . .	

where the three dots represent a model with a wholly implicit content. Models can be fleshed out to be more explicit if need be, and the initial set of models may need to contain "mental footnotes" to represent how they can be fleshed out (for details see Johnson-Laird and Byrne, 1991; Johnson-Laird, Byrne, and Schaeken, 1992).

A counterfactual conditional such as:

If Vittorio had been in Trieste then Paolo would have been in Milan.

can be represented by the fully fleshed out set of models:

factual:	not-V	not-P
counterfactual:	V	P
	not-V	P

The fully fleshed out models of the counterfactual conditional are the same as those for the factual conditional, but the models are annotated to keep track of their epistemic status. The initial set of models may leave as much information implicit as possible:

factual:	not-V	not-P
counterfactual:	V	P
	. . .	

and once again, the initial set of models may contain mental footnotes about how they can be fleshed-out to be explicit (see Johnson-Laird & Byrne, 1991). A crucial difference in the mental representation of factual and counterfactual conditionals is that the initial set of models for the counterfactual conditional is more explicit than the initial set of models for the factual conditional. The model theory proposes that counterfactual conditionals can be encompassed within the same sorts of mental representations and processes as factual conditionals, but they are represented by a richer mental representation: people keep in mind not only the false situation hypothesised by the counterfactual conditional and temporarily supposed to be true, but also the actual true situation (see also Byrne & Handley, 1997).

The theory proposes that people construct mental representations that correspond to the structure of the world rather than to the structure of the language used to describe the world. Models can represent information explicitly and implicitly, and implicit information can be fleshed out to be explicit if necessary. Models represent as little information as possible explicitly because of working memory limitations. People are rational in virtue of a semantic principle that an argument is valid only if there is no way the premises could be true and the conclusion false (Johnson-Laird and Byrne, 1991). The search for counterexamples is the central core of rationality, and the theory proposes a mental logic in which people are rational in principle but they err in practice, in part because of their limited working memory.

2.1. Inferences From Factual And Counterfactual Conditionals

The initial representation of a counterfactual conditional is more explicit than the representation of a factual conditional. Given this account of factual and counterfactual conditionals, the model theory predicts systematic differences in the inferences that people make from factual and counterfactual conditionals. We will first outline the processes by which inferences are made for one sort of inference, the *modus ponens* inference, for which we predict there will be no differences between the two sorts of conditionals. Then we will outline the processes by which inferences are made for a second sort of inference, the *modus tollens* inference, for which we predict there will be differences between the two sorts of conditionals. First consider the processes by which the *modus ponens* inference is made for the factual conditional:

If Vittorio is in Trieste then Paolo is in Milan.

Vittorio is in Trieste.

What, if anything, follows?

The first premise is represented in an initial representation:

V P

. . .

(for details see Johnson-Laird, Byrne, & Schaeken, 1992). The second premise is also represented in a model:

V

The procedures that combine models readily find a match between the two sets of models and the combination results in the elimination of the implicit models to leave only one model:

V P

which can be described parsimoniously, without repeating the premises, as:

Paolo is in Milan.

A very similar process is conducted for the following subjunctive conditional:

If Vittorio were in Trieste then Paolo would be in Milan

Vittorio is in Trieste.

What, if anything, follows?

The subjunctive conditional is represented by the following initial set of models:

factual:	not-V	not-P
hypothetical:	V	P
	. . .	

The first premise is represented in an initial set of models, and the second premise is also represented:

V

Once again the procedures that combine models leave only the model:

V P

which supports the inference:

Paolo is in Milan.

The processes are similar for the modus ponens inference from both sorts of conditionals and so we expect no difference in the ease with which people make the inference from them.

Now consider an inference, the modus tollens inference, for which the theory does predict a difference, because the initial representation contains information relevant to making the inference:

If Vittorio is in Trieste then Paolo is in Milan.

Paolo is not in Milan.

What, if anything, follows?

The first premise is represented by the initial set of models:

V P

...

The second premise is also represented:

not-P

The procedures which combine models may find no match and so reasoners may believe that nothing follows from the two premises, which is a frequent error to this inference (e.g., Evans, Newstead, & Byrne, 1993). To make the valid inference, reasoners must first flesh out the set of models to be explicit:

V P

not-V not-P

not-V P

The second premise matches the content of the second of these models and the other two models can be eliminated to leave a single model:

not-V not-P

This remaining model supports the inference:

Vittorio is not in Trieste.

The inference is difficult because it requires reasoners to flesh out the initial models, and to keep multiple models in mind. We predicted that the inference would be easier from the subjunctive conditional:

If Vittorio were in Trieste then Paolo would be in Milan

Paolo is not in Milan.

What, if anything, follows?

because the subjunctive conditional is represented by a more explicit set of models initially:

factual: not-V not-P

hypothetical: V P

. . .

The second premise is also represented:

not-P

and the two sets of models can be combined to leave the model:

not-V not-P

from which the valid inference can be made:

Vittorio is not in Trieste.

The theory predicts that the modus tollens inference will be made more often for the subjunctive conditional than the indicative one.

Of course, the rich representation constructed of a counterfactual conditional may be supplemented with background knowledge which suppresses certain inferences (e.g., Byrne, 1989a). Consider the following counterfactual (e.g., Lewis, 1973; Stalnaker, 1968):

If Hitler had had the atomic bomb then Germany would have won the war.

Reasoners represent the factual and counterfactual situations in their initial set of models:

factual:	not-H	not-G
counterfactual:	H	G
	. . .	

where "H" represents Hitler having the atomic bomb, and "G" represents Germany winning the war. If they are informed that in fact, secretly Hitler did have access to atomic weapons, they must add this information to their models, and because it matches the first part of the counterfactual model, it may seem that they should conclude that Germany won the war. But, the retrieval of background knowledge affects the fleshing-out of models, as we have argued elsewhere (e.g., in accounts of the effects of domain knowledge on Wason's selection task, Johnson-Laird & Byrne, 1991; 1996). The historical fact that Germany did not win the war ensures the inference is blocked and the truth of the counterfactual itself may be called into question.

Likewise, a counterfactual that conflicts wholly with background knowledge may be rejected in much the way that a factual conditional is. For example, the conditional:

If Hillary Clinton had been a woman then she would have been First Lady.

could be represented in the following models:

factual:	woman	first lady
counterfactual:	woman	first lady
	. . .	

and so no increase in modus tollens or the denial of the antecedent inference would be expected. But reasoners are likely to detect readily the violation of the usual conventions of counterfactual conditionals—that the factual and the counterfactual situation describe different situations—and their detection that the contents of the two models are the same may lead them to consider the counterfactual conditional to be anomalous itself.

2.2. Counterfactual Inferences

The model theory of counterfactual conditionals predicts that more modus tollens inferences will be made from a counterfactual than from a factual conditional, because of the more explicit initial representation of counterfactuals, and no differences will be observed on the modus ponens inference which can be made readily from the initial representation of even a factual conditional. We have tested these predictions of the model theory about the frequency of inferences from factual and counterfactual conditionals in a series of experiments on counterfactual conditionals (e.g., Byrne & Tasso, 1994; in press; Thompson & Byrne, 1998). In one experiment, we gave 80 undergraduates from Dublin University, Trinity College, a conditional based on facts (in the indicative mood and the present tense) or a conditional based on possibilities (in the subjunctive mood and the present tense), and we asked them what conclusion, if any, followed from the premises for four different sorts of arguments, including the two sorts we outlined earlier, modus ponens and modus tollens. As we expected, from the conditional based on possibilities, reasoners made more modus tollens inferences (80%) than they did from the conditional based on facts (40%). Also as expected, there were no differences in the frequency of modus ponens inferences from the two sorts of conditionals (100% in each case, see Byrne & Tasso, in press, for details). We examined conditionals based on impossibilities, that is, in the subjunctive mood and the past tense, as well as the present tense (Byrne & Tasso, in press), and once again the pattern supported our predictions. The results of the experiments corroborate the suggestion that the initial mental representation for counterfactual conditionals is more explicit than for factual conditionals. The more explicit representation leads reasoners to make some inferences more readily than others.

We have also examined *negative* counterfactual conditionals, such as:

If Alan had not gone to Moose Jaw then Susan would not have gone to Flin Flon.

(Thompson & Byrne, 1998). The negative counterfactual conditional may also be represented by a rich initial set of models:

factual:	A	S
counterfactual:	not-A	not-S
	. . .	

In one experiment we gave 176 Saskatchewan University undergraduates the following sort of materials:

John says that:

If Alan had not gone to Moose Jaw then Susan would not have gone to Flin Flon.

and as well as asking them about the inferences they could make, we gave them an implication task, asking them the question:

What do you think that John meant to imply? Choose as many as seem appropriate:

a) That Alan did not go to Moose Jaw

b) That Alan went to Moose Jaw

c) That Susan went to Flin Flon

d) That Susan did not go to Flin Flon

e) None of the above

The results corroborate our suggestions about which states of affairs people represent explicitly in their initial set of models for negative counterfactual conditionals. For the negative factual conditional, in so far as people thought the conditional implied a categorical proposition, they tended to believe it implied its antecedent, Alan did not go to Moose Jaw (24%), and its consequent, Susan did not go to Flin Flon (45%). For the negative counterfactual conditional, in so far as people thought the conditional implied a categorical proposition, they tended to believe it implied the negation of its antecedent, Alan went to Moose Jaw (57%), and the negation of its consequent, Susan went to Flin Flon (60%, for further details, see Thompson & Byrne, 1998). Their judgments about what a speaker intended to imply emphasize the factual situation for a negative counterfactual conditional.

3. CONCLUSIONS

We began with the question of whether it is possible to develop a general theory of conditionals to account for factual and counterfactual conditionals. The model theory proposes that people construct mental representations that correspond to the structure of the world rather than to the structure of the language used to describe the world (Johnson-Laird, 1983; Johnson-Laird and Byrne, 1991). People construct a more explicit initial representation of counterfactual than factual conditionals. As a result of these representational differences, factual and counterfactual conditionals can seem to mean something very different from each other; they can seem to support different inferences, to be verified and falsified by different states of affairs, and to imply different states of affairs. Nonetheless as we have suggested here, these differences emerge from a single cognitive mechanism: a common set of mental representations and processes that underlie reasoning about matters offact and matters of possibility and impossibility.

R.M.J. Byrne
University Of Dublin. Ireland
A. Tasso
University Of Padua. Italy
V. Thompson
University Of Saskatchewan. Canada

NOTES

* We are grateful to Phil Johnson-Laird, Mark Keane, Jamie Campbell, Vittorio Girotto, and Alberto Mazzocco for their helpful comments on the experiments.

[1] We are indebted to Phil Johnson-Laird for this example and for bringing these points about fictional worlds to our attention.

REFERENCES

Braine, M. D. S. & O' Brien, D. P., 1991. A theory of IF: a lexical entry, reasoning program, and pragmatic principles. *Psychological review, **98***, 182-203.

Byrne, R.M.J. and Tasso, A., 1994. *Counterfactual reasoning: inferences from hypothetical conditionals.* Proceedings of the sixteenth annual conference of the Cognitive Science Society. Hillsdale: Erlbaum.

Byrne, R.M.J. and Tasso, A., in press. Deductive reasoning with factual, possible and counterfactual conditionals. *Memory and Cognition..*

Byrne, R.M.J., 1989a. Suppressing valid inferences with conditionals. *Cognition, **31***, 61-83.

Byrne, R.M.J., 1989b. Everyday reasoning with conditional sequences. *Quarterly Journal of Experimental Psychology, **41A***, 141-166.

Byrne, R.M.J., 1997. Cognitive processes in counterfactual thinking about what might have been. In D.L Medin (Ed). *The Psychology of Learning and Motivation,* Vol 37. San Diego, CA: Academic Press

Byrne, R.M.J., and Handley, S.J., 1997. Reasoning strategies for suppositional deductions. *Cognition. **62***, 1-49.

Cheng, P. & Holyoak, K., 1985. Pragmatic reasoning schemas. *Cognitive Psychology. **17***, 391-416.

Cosmides, L., 1989. The logic of social exchange. *Cognition, **31***, 187-276.

Fillenbaum, S., 1974. Information amplified: memory for counterfactual conditionals. *Journal of Experimental Psychology, **102***, 44-49.

Gilovich, T. and Medvec, V.H., 1994. The temporal pattern to the experience of regret. *Journal of Personality and Social Psychology, 67*, 357-365.

Ginsberg, M. L., 1986. Counterfactuals. *Artificial Intelligence, **30***, 35-79.

Girotto, V., Mazzocco, A. and Tasso, A., 1997. The effect of premise order in conditional reasoning: a test of the mental model theory. *Cognition, **63***, 1-28.

Hofstadter, D. R., 1979. *Gödel, Escher, Bach: An Eternal Golden Braid.* New York: Basic Books.

Johnson-Laird, P.N. and Byrne, R.M.J., 1991. *Deduction.* Hove and Hillsdale: Erlbaum.

Johnson-Laird, P.N. and Byrne, R.M.J., 1996. A model point of view. *Thinking and Reasoning*, ***1***, 339-350.

Johnson-Laird, P.N. Legrenzi, P., Girotto, V., Legrenzi, M., and Caverni, J-P., 1998. Naive Probability: A mental model theory of extensional reasoning. Psychological Review. *In press.*

Johnson-Laird, P.N., 1983. *Mental Models*. Cambridge: Cambridge University Press.

Johnson-Laird, P.N., Byrne, R.M.J., and Schaeken, W., 1992. Propositional reasoning by model. *Psychological Review*, ***99***, 418-439.

Kahneman, D. and Miller, D., 1986. Norm theory: Comparing reality to its alternatives. *Psychological Review*, ***93***, 136-153.

Keane, M.T. (1997). What makes an analogy difficult ?: The effects of order and causal structure in analogical mapping. *Journal of Experimental Psychology: Language, Memory & Cognition*, ***23***, 946-967.

Landman, J., 1987. Regret and elation following action and inaction: affective responses to positive versus negative outcomes. *Personality and Social Psychology Bulletin*, ***13***, 524-536.

Lewis, D., 1973. *Counterfactuals*. Oxford, Blackwell.

Miyamoto, J. M. & Dibble, E., 1986. Counterfactual conditionals and the conjunction fallacy. *Proceedings of the Eight Annual Conference of the Cognitive Science Society*. Hillsdale: Erlbaum.

Pollock, J. L., 1986. *Subjunctive reasoning*. Dordrecht Reidel.

Ramsey, 1931. *The foundations of mathematics and other logical essays*. London: Kegan Paul.

Rips, L. J., 1994. *The Psychology of Proof*. Cambridge, MA: MIT Press.

Roese, N.J., 1994. The functional basis of counterfactual thinking. *Journal of Personality and Social Psychology*, ***66***, 805-818.

Stalnaker, R. C., 1968. A theory of conditionals. In N. Rescher (Ed.), *Studies in logical theory*. Oxford: Basil Blackwell.

Thompson V. A. and Byrne, R.M.J., 1998. Negative counterfactual conditionals. Manuscript in preparation.

Thompson V. A., 1994. Interpretational factors in conditional reasoning *Memory & Cognition*, ***22***, 742-58

Thompson V. A., 1995. Conditional reasoning: The necessary and sufficient conditions. *Canadian Journal of Experimental Psychology*, ***49***, 1-60

STEPHEN SCHIFFER

MEANINGS AND CONCEPTS

1. PROPOSITIONS

Sentences, speech acts, and thoughts are alike in that they have *propositional content*. Thus, 'La neige est blanche' means *that snow is white*; in uttering 'Over my dead body', Betty was letting you know *that the probability of her going out with you wasn't very high*; and one of your mental states is a belief *that Palermo is south of Rome*. Because sentences, speech acts, and thoughts all have propositional content, one can't sensibly limit one's semantic interests to the philosophy of language; the theory of content, my concern in this paper, is defined by issues that cut across both the philosophy of language and the philosophy of mind.

As we all know, philosophical investigations can't be neatly contained within the traditional curriculum headings of philosophy of language, epistemology, metaphysics, and so on, and one of the things that makes the theory of content exciting is how quickly it leads to some hairy issues in ontology. For consideration of sentences that ascribe propositional content has led Frege and many others to suppose that the truth of these sentences requires the existence of abstract *entities* called 'propositions', and many other philosophers have found propositions so mysterious or otherwise repugnant that they've gone to great lengths to try to avoid being committed to their existence. Quine once got so lathered about propositions that he called them creatures of darkness (if Quine were more widely read there would no doubt be a cult rock band called *The Propositions*). Still, there is a case for the existence of propositions, and we may put it in the following way. It's a two-part case, and it's first part is a case for the claim that that-clauses, such as 'that Fido is a dog' in 'Ralph believes that Fido is a dog', are *referential singular terms*, where to say that an expression t in a sentence $S(t)$ is a referential singular term is to say, at least to a rough first approximation, that t stands for an object x such that the sentence $S(t)$ is true just in case the predicate $S(\ )$ is true of x. So, for example, we say that 'Fido' in 'Fido is a dog' is a referential singular term because there is a certain dog, Fido, such that the sentence is true just in case the predicate 'is a dog' is true of Fido, and we say that 'the woman over there' in an utterance of 'Lester loves the woman over there' is a referential singular term because that utterance is true just in case the predicate 'Lester loves _' is true of the indicated woman. Now, the reason for taking that-clauses to be referential singular terms is

K. Korta et al. (eds.), Cognition, Agency and Rationality, 79-92.

quite simply that it's apparently the best way to account for the evident validity of such inferences as the following:

> Lester believes that eating liver increases sexual potency, and so does Carlotta.
> So, there is something that they both believe—to wit, that eating liver increases sexual potency.
>
> Lester believes that eating liver increases sexual potency.
> That eating liver increases sexual potency is Carlotta's theory.
> So, Lester believes Carlotta's theory.
>
> Lester believes everything that Carlotta says.
> Carlotta says that eating liver increases sexual potency.
> So, Lester believes that eating liver increases sexual potency.

We can readily account for the validity of these inferences if we suppose that the that-clause contained in them, 'that eating liver increases sexual potency', is a referential singular term, and it's not clear that we can account for it otherwise.[1] In connection with this, notice how in the premise 'That eating liver increases sexual potency is Carlotta's theory' the that-clause occupies the grammatical subject position, and how we can substitute the singular term 'Carlotta's theory' for it in the sentence 'Lester believes that eating liver increases sexual potency'. Despite this evidence, there do exist various attempts to account for the validity, or seeming validity, of these inferences without the assumption that that-clauses are referential singular terms. There's even my own labored attempt in my book *Remnants of Meaning*[2] to use non-objectual quantification in aid of the denial project. My present view, however, is that all these attempts fail; that-clauses really do typically function as referential singular terms.[3] At any rate, that they do so function will be a working assumption of the rest of this paper.

Very well, that-clauses refer. The next question, then—turning to the second part of the two-part case for propositions—is: To what do they refer? What are the referents of that-clauses and, thereby, the things we believe and assert? That is easy to answer: the referent of 'that eating liver increases sexual potency' is *that eating liver increases sexual potency*; *that eating liver increases sexual potency* is precisely the referent of the that-clause singular term. To be sure, to be sure, I can hear you muttering, but what manner of thing is this thing, *that eating liver increases sexual potency*, which is the referent of the that-clause singular term? Happily, there are a number of things we can say in response to this question right off the tops of our heads. First, *that eating liver increases sexual potency* is *abstract*, or *immaterial*. It doesn't occupy space and has no physical properties at all. Second, it's *mind and language independent* in at least two senses: it exists in possible worlds in which there are neither thinkers nor speakers, and although it can be expressed by a sentence of every language, it itself belongs to no language; *that eating liver increases sexual potency* isn't Japanese, Italian, or English. Third, it has a *truth condition*: *that eating liver increases sexual potency* is true iff eating liver increases sexual potency. Fourth, it has its truth condition *essentially*; it's a *necessary truth* that *that eating liver increases sexual potency* is true iff eating liver increases sexual potency. This is in contrast to the *sentence* 'Eating liver increases sexual potency', which, while also true iff eating liver increases sexual potency, has its truth condition only contingently on our actual linguistic practices. Had our use of language been different, it might have had a different truth condition or none at all. Fifth, and finally, *that eating liver increases sexual potency* has its truth condition *absolutely*, without relativization to anything. The contrast is again with the sentence 'Eating liver increases sexual potency', which has its truth condition only *in English* or *among us*, and may have a different truth condition in some other language or among some other population of speakers. But *that eating liver increases sexual potency* has its truth condition everywhere and everywhen. In short, the referents of that-clauses, and therewith the contents of our speech acts and propositional attitudes, are what philosophers call *propositions*: abstract, mind- and language-independent entities that have truth conditions, and have their truth conditions both essentially and absolutely.

As is well known, there are those who agree that that-clauses refer but who deny that they refer to propositions. These guys attempt to make do with linguistic surrogates, and I can here say only that in my opinion they fail. At any rate, another working hypothesis of this paper will be that propositional attitudes really are relations to propositions, to those propositions to which that-clauses refer. If we're stuck with propositions, then we should try to demystify their existence, and I'll have something to say about this later.

2. THE FREGEAN THEORY OF PROPOSITIONS

Our working hypothesis is that propositions, being the referents of that-clauses, are the things we mean and believe. This still leaves plenty of room for propositionalists to disagree among themselves about what else is true of propositions. It is arguable that the most plausible further account of propositions remains Frege's. At all events, the Fregean position is still a dominant position, and it enjoys its most recent important development in Christopher Peacocke's *A Study of Concepts*.[4] The Fregean position may be characterized in the following way.

Pretend that the Superman fiction is fact and consider these two sentences:

Lois Lane believes that Superman eats groundhogs.

Lois believes that Clark Kent eats woodchucks.

The Fregean, quite sensibly, holds two things initially. First, she holds that these sentences may differ in truth-value notwithstanding that Superman = Clark Kent and the property of being a groundhog = the property of being a woodchuck. Second, she holds that these sentences have the form they appear to have: 'believes' occurs in them as standing for a two-place relation holding between believers and the propositions they believe, while the two singular terms in each sentence stand for alleged terms of that relation.[5] In other words, the Fregean would claim that these two sentences enjoy the following form-revealing representations:

B(Lois, the proposition that Superman eats groundhogs)

B(Lois, the proposition that Clark Kent eats woodchucks)

These two sensible initial assumptions commit the Fregean to a simple and plausible account of how it is that our two belief sentences can differ in truth-value—viz., the two that-clauses refer to distinct propositions. The proposition that Superman eats groundhogs is not the same proposition as the proposition that Clark Kent eats woodchucks.

The Fregean next explains why these two propositions are different: they have different *constituents*. Propositions, for the Fregean, are structured entities whose basic constituents are what we may call *propositional building blocks* (although, I warn you, I'm about to call them three other things as well!). Our two propositions are different because they're built from different building blocks. These propositional building blocks are *the references words have in that-clauses*. Let me explain. Some singular terms are semantically simple in that their references aren't determined by the references or semantic values of any of their parts. Proper names, such as 'Fido', are like that. Other singular terms are semantically complex in that their references are determined by their syntax and the references of their constituent expressions. 'The capital of Italy' owes its reference to its syntax and the fact that 'Italy' refers to Italy and 'the capital of' refers to that function which maps countries onto their capitals. Evidently, that-clauses are semantically complex singular terms, given that they are singular terms. According to the Fregean, propositional building blocks are the references words have in that-clauses (which references, we can already deduce, won't be references words have outside of that-clauses).

We still haven't asked what sorts of things the Fregean takes propositional building blocks—the references words have in that-clauses—to be, but before turning to that important question, let's notice three further things we can say about Fregean propositional building blocks. The first is a merely verbal point, but verbal points can be interesting. The word 'concept' is used in philosophy as a term of art, although it's unfortunately used as more than one term of art. But a dominant use, highly congenial to the Fregean, is that concepts are constituents of the contents of thoughts. On this common way of speaking, propositional building blocks, whatever they turn out to be, are *concepts*. The second further thing I wanted to say about Fregean propositional building blocks is that, subject to a certain qualification, they're also *word meanings*. We get this result in the following simple way. Consider the true sentence

'Superman eats groundhogs' means that Superman eats groundhogs.

Now, it's a platitude that the meaning of a word is its contribution to the meanings of the sentences in which it occurs, and, as the displayed truth illustrates, the contribution that the words in 'Superman eats groundhogs' contribute to the meaning of the sentence are precisely the references those words have in the displayed that-clause. The qualification to which I alluded has to do with the fact that we can't regard every indicative sentence as meaning a complete proposition (e.g., the sentence type 'She's now there' expresses no complete proposition). But it won't hurt us to ignore this complication for now and to appreciate what for the Fregean is the near truth that propositional building blocks are both concepts and meanings.

The third thing I wanted to say at this point about Fregean propositional building blocks is that, for the Fregean, the truth-value of a proposition is determined in a certain way by "semantic values" of its constituent concepts. We've already noticed that for the Fregean propositional building blocks can't be the objects and properties our beliefs are about. If they were, then the proposition that Superman eats groundhogs would be identical to the proposition that Clark Kent eats woodchucks. Still, propositional building blocks must bear *some* important relation to those ordinary objects and properties, otherwise there would be nothing to make our beliefs about those things. The Fregean suggests an accommodation of this constraint via her use of the metaphor of a *mode of presentation*. Before proceeding, though, let's pause to keep from getting overwhelmed by a surfeit of labels. According to the Fregean, the propositions we believe and assert are structured entities, and we're already calling their constituents *propositional building blocks*, *concepts*, and *meanings*. The Fregean also calls them *modes of presentation*. Her point is that while propositional building blocks aren't the objects and properties our beliefs are about, they are *modes of presentation* of those things. Modes of presentation are propositional building blocks, and the things of which they are modes of presentation are their "semantic values." Use of the metaphor of a mode of presentation affords a neat way of spelling out what the Fregean means in saying that the truth-value of a proposition is determined by, or is a function of, semantic values of its constituent concepts. At a certain level of analysis, the Fregean can represent all propositions as being of the form

$$<<m_1,\ldots, m_n>, m^n>$$

where $<m_1,\ldots, m_n>$ is an *n*-ary sequence of modes of presentation of things of any kind and m^n is a mode of presentation of an *n*-ary relation (one-place relations are properties). Then the sense in which, for the Fregean, the truth-value of a proposition is determined by semantic values of its constituent concepts is given by the following definition of truth and falsity for propositions ('$\exists!v$' means "there is a *unique* v such that"):

$$<<m_1,\ldots, m_n>, m^n> \text{ is}$$

true iff $\exists!x_1,\ldots,x_n\exists!\phi^n(m_1,\ldots, m_n$ are modes of presentation of $x_1,\ldots, x_n$ respectively & m^n is a mode of presentation of ϕ^n & $<x_1,\ldots, x_n>$ instantiates $\phi^n)$;

false iff $\exists! x_1,\ldots,x_n \exists! \phi^n(m_1,\ldots, m_n$ are modes of presentation of $x_1,\ldots, x_n$ respectively & m^n is a mode of presentation of ϕ^n & $<x_1,\ldots, x_n>$ doesn't instantiate $\phi^n)$;

neither true nor false iff $\sim\exists! x_1,\ldots,x_n \exists! \phi^n(m_1,\ldots, m_n$ are modes of presentation of $x_1,\ldots, x_n$ respectively & m^n is a mode of presentation of $\phi^n)$

Thus, the proposition that Fido is a dog may be represented as

$$<m_f, m_d>$$

where m_f is a mode of presentation of Fido and m_d is a mode of presentation of doghood (and where we allow ourselves to drop the brackets for unit sequences). This proposition is therefore true just in case Fido instantiates doghood, i.e., just in case Fido is a dog. The proposition that Fido loves Gina may be represented as

$$<<m_f, m_g>, m_l>$$

where m_f, m_g, and m_l are modes of presentation respectively of Fido, Gina, and the love relation, and the proposition is therefore true just in case <Fido, Gina> instantiates the love relation; i.e., just in case Fido loves Gina. And the complex proposition that roses are red and violates are blue may be represented as

$$<<\mathrm{m_r}, \mathrm{m_v}>, \mathrm{m_{conj}}>$$

where m_r, m_v, and m_{conj} are modes of presentation respectively of the proposition that roses are red, the proposition that violets are blue, and the conjunction relation, and the proposition is therefore true just in case <the proposition that roses are red, the proposition that violets are blue> instantiate the conjunction relation; i.e., just in case roses are red and violets are blue.

3. MODES OF PRESENTATION AS PLEONASTIC CONCEPTS

This brings us to the $64 question: What *are* Fregean modes of presentation? What sorts of things satisfy the foregoing characterization of modes of presentation? As so far characterized, our understanding of the notion of a mode of presentation is simply that they are whatever things play such-and-such theoretical role, if indeed there are things that play that role. What I'm now asking is *what things*, if any, play that role. To ask this question, we know, is the same as to ask what, for the Fregean, are *concepts* or *meanings*.

One strategy for answering this question would be to try to give an account of modes of presentation which satisfies what I've elsewhere called the *intrinsic-description constraint*.[6] According to this constraint, if a thing is a mode of presentation—if, that is, it plays the mode-of-presentation role—then it must be intrinsically identifiable in a way that does not describe it as a mode of presentation or as a possible mode of presentation. If a thing is a mode of presentation, then it must be intrinsically identifiable as some other kind of thing. Denying this constraint is apt to seem tantamount to introducing the notion of a gene as whatever plays such-and-such role in the transmission of inheritable characteristics and then insisting that the things which play that role enjoy no more intrinsic characterization than 'things that play such-and-such role in the transmission of inheritable characteristics'. To deny the intrinsic-description constraint is to insist that propositional building blocks enjoy no more intrinsic characterization than that they are propositional building blocks. So the intrinsic-description constraint is hardly unmotivated. At the same time, I'm pretty confident that the Fregean won't be able to satisfy the constraint. As I've argued elsewhere,[7] when you go through the list of candidates for modes of presentation that satisfy the constraint, you can find pretty good reasons for striking each candidate off the list. I can't review the case for this now, but let me briefly mention one example. Many have read Frege, rightly or wrongly, as suggesting that modes of presentation are *uniqueness properties* of the form *the property of uniquely having the property* ϕ—that is, that property a thing has when it has the property ϕ and nothing else has ϕ. Whether or not Frege actually held this, it's clear that Bertrand Russell once did. But the view is hopeless. Since properties need modes of presentation, you couldn't think of something under a mode of presentation without having a distinct mode of presentation for that mode of presentation, thereby setting off a self-refuting regress.

If the Fregean theory is to have a chance, it must deny the intrinsic-description constraint. This is recognized by the Fregean Christopher Peacocke, who in his *A Study of Concepts* offers an account of concepts—i.e., modes of presentation—which doesn't satisfy the constraint, and then explicitly argues that "the intrinsic-description constraint is quite generally false of abstract objects, and its falsity for concepts or modes of presentation is a special case of this general falsity."[8] I'm now inclined to agree with Peacocke that there is a reasonable way of denying the intrinsic-description constraint, although this way isn't exactly Peacocke's and although I believe that the theory of modes of presentation Peacocke offers is less than fully correct. But this isn't the occasion to discuss Peacocke's interesting views on these matters, and in the remainder of this paper I'm going to lay out, as simply and as baldly as I can, what I think must be said if one's to have modes of presentation that don't satisfy the intrinsic-description constraint. The Fregean, I'll propose, needs to identify modes of presentation, her propositional constituents, with what I'm soon to call *pleonastic concepts*. Let me explain.

It's my view that certain kinds of objects are in a sense language-created "linguistic posits," hypostatizations of certain linguistic practices, even though, in another sense, they enjoy a mind- and language-independent existence.[9] Propositions are linguistic posits in this sense. They are mind and language independent in the two senses already mentioned: propositions would have existed no matter what linguistic or conceptual practices we employed, and one and the same proposition can be grasped and expressed by speakers of different languages. They are linguistic posits, hypostatizations of the linguistic practices that introduce propositions into our ontology, in a sense that includes the following claims.

1. Linguistic posits typically arise from what may be called *something-from-nothing transformations*. These are trivial transformations that take one

from a sentence in which no singular term refers to the linguistic posit to a sentence that does contain such a singular term. Thus, from

Fido is a dog,

whose only singular term is 'Fido', we can infer its pleonastic equivalent

That Fido is a dog is true,

or, more colloquially,

It's true that Fido is a dog,

which contains the singular term 'that Fido is a dog' whose referent is the proposition that Fido is a dog. It's because of our ability to move back and forth between any sentence '*S*' and its pleonastic equivalent 'That *S* is true' that we have the well-known truth schema for propositions:

(The proposition) that *S* is true iff *S*.

(The existence of something-from-nothing transformations doesn't imply that all references to linguistic posits can be paraphrased away. There's no paraphrasing away the that-clause in 'Ralph believes that Fido is a dog'. But the use of that-clauses in these constructions is parasitic on its uses in the hypostatizing something-from-nothing transformations.)

2. To have knowledge of linguistic posits, one merely needs to be party to the linguistic practices by which they are introduced, and there is no other way of gaining knowledge of them. Imagine a possible world, β, exactly like the actual world, α, except that in β we don't have linguistic practices that license the formation of that-clauses; we in that world lack the practice that allows us to transform the sentence '*S*' into the singular term 'that *S*'. It follows from our actual practices governing that-clauses, our practices in α, that propositions exist in β, but, lacking the concept of a proposition, we in β would lack all knowledge of them. What would it take to bring us in β up to epistemological snuff with us in α? It's simple: what we'd need to do, and all that we'd need to do, is adopt the proposition-introducing language games we actually play. We certainly couldn't become aware of the existence of things that are *not* linguistic posits in this way. You couldn't, for example, become aware of *trees* simply by introducing talk of trees. You would *first* have to discover trees and *then* introduce talk of trees. An important corollary of this point is that no substantial relation—certainly no causal relation—must obtain between us and propositions in order for us to refer to them and to have knowledge about them; it's enough that we indulge in a certain linguistic practice, a certain language game.

3. There is nothing more to the *nature* of linguistic posits than is determined by the hypostatizing linguistic practices by which linguistic posits are

introduced. What we can learn about them is what our linguistic practices license us to learn about them. The essences of things that enjoy the highest degree of independence from our linguistic and conceptual practices—trees and electrons, for example—can be discovered by a posteriori, scientific investigation, but the essence of linguistic posits can't be discovered in any such way. Whatever belongs to their essence can be read off the something-from-nothing linguistic practices that posit them in our ontology. As Mark Johnston aptly puts it, linguistic posits have no "hidden and substantial nature for a theory to uncover."[10]

4. A corollary of the nature-determination point, but one worth mentioning separately, concerns the individuation of linguistic posits.[11] Linguistic posits needn't have nontrivial criteria of individuation, and this is true of propositions. At the same time, we are able to make confident judgments of nonidentity, for we have criteria for ascribing properties to propositions. Just think of all the differences between the proposition that Harry Truman had toes and the proposition that the Pope is Catholic. In a sense, what "makes" the proposition that Superman eats groundhogs distinct from the proposition that Clark Kent eats woodchucks is that someone can believe one of them without believing the other. I'm not trying to say something trivial. If Al kissed Betty but did not kiss Carla, then Betty ≠ Carla, and if Lois believes that Superman eats groundhogs but doesn't believe that Clark Kent eats woodchucks, then the proposition that Superman eats groundhogs ≠ the proposition that Clark Kent eats woodchucks. However, the important difference between the two kinds of cases is as follows. Suppose you were wondering whether Betty and Carla were the same person. You could hardly hope to determine that they were not by determining that Al kissed Betty but not Carla. Imagine saying, "Well, I saw that Al kissed Betty and that he didn't kiss Carla, so I knew Betty wasn't Carla." The absurdity, of course, is that you couldn't rationally conclude in this case that Al kissed Betty but not Carla unless you already knew that Betty was not Carla. But you *can* determine that Lois believes that Superman eats groundhogs and that she doesn't believe that Clark Kent eats woodchucks without any prior opinion, as it were, about the identity or difference of the two propositions. It's because the criteria for the truth of these belief statements are independent in this way that we can *first* determine the different truth-values of 'Lois believes that Superman eats groundhogs' and 'Lois believes that Clark Kent eats woodchucks' and *then* conclude that the proposition that Superman eats groundhogs ≠ the proposition that Clark Kent eats woodchucks. In short, our criteria for determining the truth of belief statements doesn't require a prior individuation of the propositions involved; rather, our individuation of propositions is culled entirely from prior criteria for assessing belief statements.

What I now want tentatively to propose is that we might construe ourselves as having linguistic practices that afford us a conception of concepts as linguistic posits in pretty much the way propositions are. *Pleonastic propositions* is what I like to call propositions conceived as linguistic posits, so the view of concepts I'm proposing may be called a conception of *pleonastic concepts*. The primary linguistic practice I have in mind is the one that licenses inferences like the following:

> Giorgio believes that Satan lurks everywhere.
> So, Giorgio believes something involving his concept of Satan.

To be sure, we don't seem to have very rich practices involving the word 'concept', but the foregoing doesn't seem too strained, and is a version of a something-from-nothing introductory practice in that 'Giorgio believes that Satan lurks everywhere' contains no singular term that explicitly refers to the concept of Satan.[12] Given that we have such a practice, concepts will be linguistic posits in just the way propositions are. Among other things, this means that our only basis for knowing that the concept of $X \neq$ the concept of Y is that one can believe that ... X ... without believing that ... Y .... Criteria for ascribing beliefs come first, and from them we cull our ways of individuating concepts. Better yet, criteria for ascribing beliefs come first, and from them we cull our ways of individuating propositions, and from them we cull our ways of individuating concepts.

There's more that we can say about this deflationary conception of concepts that we find riding piggyback on our deflationary conception of propositions, although I'm unable here to give a complete elaboration.

a. Pleonastic concepts, like the pleonastic propositions from which they derive, don't satisfy the intrinsic-description constraint. The only intrinsic characterization of the concept of Satan is 'the concept of Satan'.

b. As we'll presently observe, the context 'the concept of ...' may be construed as admitting of two readings, but the one that's been in play so far is intentional in two senses. First, the concept of X may exist even though X doesn't exist. Giorgio's concept of Satan exists even though, presumably, Satan doesn't exist. Second, it may be that the concept of $X \neq$ the concept of Y even though $X = Y$. For example, even though Superman = Clark Kent, Lois's concept of Superman $\neq$ her concept of Clark Kent.

c. We noticed that for the Fregean propositional building blocks must determine semantic values that can be used to determine the truth-values of the propositions those building blocks help to build. A major task for the Fregean is to give a theory of this determination, a theory that tells us how the constituents of propositions determine their semantic values. When the Fregean calls these building blocks *concepts*, the task is to say what it is for a particular object or property to fall under a concept. Our pleonastic conception of concepts affords us an easy way of characterizing the determination relation. We've noticed that since the position of 'X' in 'the concept of X' is intentional (at least as so far characterized), we can't say that a particular object or property falls under the concept of X just in case that object or property $= X$. However, we can say that every instance of the schema

> If X exists, then X falls under the concept of X

is analytic in just the way that every instance of the schema

The proposition that S is true iff S

is analytic. These schemas are trivial consequences of our ways of introducing talk of concepts and propositions respectively. Since we can speak of a thing as falling under the concept of X, we may easily introduce a nonintentional way of understanding 'a concept of y', the lower-case letter being our way of marking the difference. We can say that, for any y, the concept of X is a concept of y just in case y falls under the concept of X.

d. The way pleonastic concepts are obtained from that-clauses suggests we can construe them as the references words have in that-clauses, for each word in a that-clause determines a concept involved in the proposition to which the whole that-clause refers.

In this way we arrive at a singularly deflationary version of Frege's theory of propositions which, since I'm already speaking of pleonastic propositions and concepts, we might as well call *pleonastic Fregeanism*. It's the view that the propositions we believe and assert are pleonastic propositions composed of pleonastic concepts. It's the view that Fregean modes of presentation are pleonastic concepts. One reason that this is not full-blown Fregeanism is that for Frege what *explains* the fact that the proposition that Superman eats groundhogs ≠ the proposition that Clark Kent eats woodchucks is that they have different constituents. But on our notion of pleonastic concepts, the fact that these two propositions have different constituents is entirely derivative on their being different propositions, which in turn, I've suggested, is entirely derivative on the conceptually prior fact that 'Lois believes that Superman eats groundhogs' and 'Lois believes that Clark Kent eats woodchucks' may differ in truth value.

4. PLEONASTIC FREGEANISM AND COMPOSITIONAL SEMANTICS

I now want to conclude by bringing pleonastic Fregeanism to bear on the vexing question of compositional semantics. For simplicity, let's ignore indexicality, ambiguity, vagueness, and grammatical moods other than the indicative. Then we may say that:

> A *compositional meaning theory for a language L* is a finitely statable theory of L that associates compositional rules with the basic syntactic structures of L and assigns meanings to the primitive words of L in such a way as to generate for each sentence of L a truth of the form
>
> σ means p in L.

Thus, a compositional meaning theory for French stated in English would entail the statement

'La neige est blanche' means in French that snow is white.

If we assume, with Frege, that sentence meanings are structured propositions, then the meanings assigned to the words of *L* will be propositional building blocks. Now, it's very widely held that each natural language has a compositional meaning theory. It's held that each natural language has such a compositional semantics because the hypothesis that it does is needed to explain certain things. Among the things theorists have claimed we need a compositional meaning theory to explain are:

(i) Our ability to understand utterances of novel sentences, sentences we've never heard before. It's argued that what explains this is that although the sentence was novel, its words and structure weren't. We could figure out the meaning ofthe novel sentence because we already knew the meanings of its parts and knew a rule for putting those meanings together to get the meaning ofthe sentence. And this story, it's further held, makes sense only on the assumption that languages have compositional meaning theories.

(ii) The productivity oflanguage: the fact that each ofinfinitely many sentences has its own unique meaning.

(iii) The systematicity of language: the fact that each word makes a uniform contribution to the meanings of the infinitely many sentences in which it occurs.

(iv) The productivity ofthought: our ability in principle to entertain any one of an infinity of propositions.

(v) The systematicity of thought: the fact that the ability to entertain any one thought carries with it the ability to entertain numerous permutations of that thought. For example, someone who can entertain the thought that John loves Mary can also entertain the thought that Mary loves John.

For the past ten years, I have been arguing that languages neither have nor need compositional semantics. I now think I may have been only half right. If we avail ourselves of pleonastic concepts as word meanings, then we can allow that languages have compositional meaning theories—*pleonastic* compositional meaning theories. So I may have been wrong to think that languages don't have compositional meaning theories. But pleonastic compositional meaning theories won't *explain* any of the things theorists have thought they needed compositional semantics in order to explain. This is because of the way pleonastic concepts are abstractions from the already determined pleonastic propositions and not genuine building blocks of them. As I've argued elsewhere,[13] the issue about whether languages have compositional meaning theories boils to down the issue of whether propositions are compositionally determined. Propositions are compositionally determined if there's a finitely definable function from sequences of propositional building blocks onto the propositions they build. If there are pleonastic concepts, then propositions are compositionally determined. But only in a very Pickwickian sense, because there's no identifying the building blocks until you already have the propositions they build. Pleonastic concepts are an epiphenomenon of that-clauses, and they contribute nothing to the mechanisms whereby that-clauses determine propositions. Pleonastic compositional meaning theories would explain what needs to be explained only if pleonastic concepts really were an essential part of the mechanism that explained the business ofthat-clauses.

Well, if pleonastic compositional semantics is the best compositional semantics we can have, and if it doesn't explain language understanding and the different versions of productivity and systematicity, then what does explain those things? Our understanding of natural languages may be quite easy to explain without a compositional semantics: as I argued in Remnants of Meaning, we don't need a compositional semantics to explain our understanding of Mentalese, our internal system of mental representation, and natural language understanding can be understood wholly in terms of certain "translation" functions that map spoken utterances onto meaning-equivalent Mentalese sentences but do so wholly on the basis of the syntactic features of the sentences on which they operate.[14] We would still need to explain the productivity and systematicity of Mentalese, and therewith, directly or indirectly, the productivity and systematicity of thought and natural languages, but most of that can be done via what I've elsewhere called compositional supervenience theories,[15] theories that are compositional but don't imply compositional semantics. And so it goes, one thing leading to another, and where will it ever end? Not here, not now.

Stephen Schiffer
New York University
U.S.A.

NOTES

[1] For present purposes, I count that-clauses as referential singular terms even if they're to be analyzed on an analogy with Russell's treatment of primary occurrences of definite descriptions.
[2] MIT Press (1987).
[3] They don't quite function in this way when they're being quantified into, as in 'Mary believes of some student that he plagiarized his paper'.
[4] MIT Press (1992).
[5] I ignore tense and temporal reference for simplicity.
[6] See my "The Mode-of-Presentation Problem," in C. Anderson and J. Owens, eds., *Propositional Attitudes* (Stanford: CSLI (1990)), and "Belief Ascription," *The Journal of Philosophy*, LXXXIX, 10 (October 1992): 499- 521.
[7] Schiffer, op. cit.
[8] Ibid., p. 121.
[9] The position now to be sketched without a supported defense is more fully elaborated, and defended, in my "Language-Created Language-Independent Entities," *Philosophical Topics* 24 (1) (1996): 149-167. See also my "A Paradox of Meaning," *Noûs* 28 (1994): 279-324, Mark Johnston, "The End of the Theory of Meaning," *Mind & Language* 3 (1988): 153-185, and Robert Stalnaker, "On What Possible Worlds Could Not Be," in A. Morton and S. Stich, eds., *Benacerraf and His Critics* (Blackwell (1996)).
[10] Johnston, op. cit., p. 38.
[11] What follows is culled from "A Paradox of Meaning," pp. 311-313.
[12] Alex Barber, in "The Pleonasticity of Talk About Concepts," forthcoming in *Philosophical Studies*, also develops a theory of what he, too, calls pleonastic concepts. Although our two conceptions of pleonastic concepts aren't the same, there are, as you'd

expect, important affinities, and I especially applaud his implying that concepts are mere epiphenomena of the something-from-nothing linguistic transformations by which they're introduced.

[13] "A Paradox of Meaning."

[14] See also Jerry Fodor, "Review of Stephen Schiffer's *Remnants of Meaning*," *A Theory of Content and Other Essays* (MIT Press (1990)).

[15] See especially "A Paradox of Meaning."

THOMAS HOFWEBER

CONTEXTUALISM AND THE MEANING-INTENTION PROBLEM

1. CONTEXTUALISM AND SEMANTIC CONTEXT DEPENDENCE

The relevant alternatives approach in epistemology[1] arose some years ago partly out of the hope to be able to reconcile our ordinary claims of knowledge with our inability to answer the skeptic. It was supposed to give rise to an account of knowledge according to which our ordinary claims of knowledge are true, even though the claims about our lack of knowledge that the skeptics make in one of their more persuasive moments are also true. To know, according to such an account, is to have evidence sufficient to rule out all the relevant alternatives. In ordinary life few alternatives are relevant. For example, whether or not we are brains in a vat is not a relevant alternative that we have to be able to rule out. In the debate with the skeptic it may become relevant, and accordingly we might not know something any more then, even though we have the same evidence as in ordinary life. The skeptics cleverly make more and more alternatives relevant, and that is how they succeed. But their success in the philosophy seminar is no threat to our ordinary claims of knowledge, or so the theory goes.

Originally the formulations of relevant alternatives theories have not been very explicit about whether or not this theory should be understood as a theory about the content of knowledge ascriptions, or as a theory stating conditions for whether or not a person knows a proposition in certain circumstances. In the first case one would have to spell out under what conditions an ascription of knowledge in a certain context is true. In the second case one would have to spell out under what conditions a person knows a proposition in a certain context. This might not seem a big difference, but once we look more closely at what role the context plays in these two we will see a difference. In the first the context is the context of the utterance of the ascription of knowledge, in the second it is the context of the alleged knower. So what contextual features will be relevant will be quite different in the two cases.[2]

In more recent years, several philosophers[3] have developed the relevant alternatives approach to knowledge into a theory that is very clear and explicit about which of these two formulations it should take. According to these philosophers this set of ideas should be understood as a theory about the content of knowledge ascriptions, as a theory about what is said with an ascription of knowledge in a certain context. It is claimed that the content of a knowledge ascription can change

K. Korta et al. (eds.), Cognition, Agency and Rationality, 93-104.

from context to context, and it is because of this that our ordinary claims to knowledge and the claims the skeptic makes are compatible. Because in ordinary live and in the debate about skepticism we are in contexts that are in important respects different, what we say with our ascriptions of knowledge and of lack of knowledge are compatible. Simply put, one speaks truly when one says "I know that I have hands" in a context where the low standards for knowledge of everyday life are relevant, but falsely in a context where the high standards of knowledge in the debate about skepticism are relevant. A position of this kind is called *contextualism* (to be more precise one should call it a contextualism about knowledge ascriptions).

Thus a contextualist claims that knowledge ascriptions are context sensitive in the sense that what is said with such an ascription depends on the context of its utterance. Knowledge ascriptions are thus a case of semantic context sensitivity. In general, a sentence is semantically context sensitive if the content of an utterance of it depends on the context in which it was uttered. A term is semantically context sensitive if the contribution it makes to the content of an utterance of a sentence in which it occurs depends on the context of this utterance. Semantic context sensitivity is a well known linguistic phenomenon, and there are a number of well described and investigated cases of it. Most obviously, there is the case of indexicals, like "I". It contributes the speaker of the utterance to the context of the utterance. Then there is the case of ambiguity and semantic underdetermination. What is said with an utterance of a certain sentence depends in these cases on other much more complicated features of the context. Consider examples like:

(1) Three girls wrote four letters.

(2) John's book is blue.

For (1) it will have to get determined in the context of utterance what exactly the truth conditions of an utterance of it are. Is it claimed that each of the three girls wrote four letters by herself, or did they write four letters together? For (2) the context will have to fill in what relation is supposed to hold between John and a certain book. Is he the writer, the owner, the one who is holding it, etc.?

These cases, however, do not seem to have much to do with knowledge ascriptions, at least on the face of it. What seems to be much closer to the case of knowledge ascriptions, and what is usually used by contextualists as an example to motivate contextualism, are expressions like "tall" or "flat". These are semantically context sensitive because something is not tall or flat *simpliciter* but only tall or flat relative to a comparison class. A sentence like

(3) Joe is tall.

can be uttered to express different propositions. What the relevant comparison class is will differ from context to context. The same sentence with reference to the same Joe, can be uttered truly when talking about 12 year olds, or falsely when talking about basketball players.

To understand this case better and how it relates to the case of knowledge ascription I would like to bring in some terminology. Lets, for the sake of a distinction to be made, take propositions to be modeled by structured entities, like n-tuples, that contain constituents. Thus the proposition that Joe is male will be modeled by the pair consisting of Joe and the property of being male as its only two constituents. Given this model of propositions, we can distinguish between articulated constituents and unarticulated constituents.[4] A constituent of a proposition expressed by an utterance of a certain sentence is articulated if there is a phrase in the sentence uttered that contributes that constituent to the proposition expressed. For example, the proposition expressed by an utterance of "Joe is male" has only articulated constituents, since Joe is contributed to the proposition by "Joe", the property of being male is contributed by the phrase "is male", and there are no other constituents. However, the proposition expressed by an utterance of "Joe is tall" will contain an unarticulated constituent, since it will express the proposition that Joe is tall for an F. The proposition expressed by this utterance will contain the class of Fs as a constituent, even though there is no phrase in the sentence uttered that contributes that constituent to the proposition expressed. So, in the case of "Joe is tall" it will be such that the proposition expressed will contain a comparison class as an unarticulated constituent, and which one this is depends on the context of the utterance.

We can distinguish between two cases of unarticulated constituents. These differ in the cognitive relation that the utterer has towards the unarticulated constituent that is contributed to the proposition expressed. We will say that we are dealing with a case of <u>implicit relativity</u> if the proposition express with an utterance of a certain sentence contains an unarticulated constituent and the speaker has cognitive access to the fact that this is so and what this constituent is. This can be tested simply by asking the speaker relative to what they meant to say that X. This will occur in the case of an utterance of "Joe is tall". When we ask the speaker relative to what they meant to say that Joe is tall then they will rightly answer that they meant to say that Joe is tall for a basketball player, or for a 12 year old. This first case of relativity certainly occurs and seems to be quite common. The other case is much more problematic and controversial, as we shall see. We shall say that we are dealing with a case of <u>hidden relativity</u> when the proposition expressed by an utterance of a sentence contains an unarticulated constituent, and the speaker of that sentence has no cognitive access to the fact that this is so. In this case the speaker will simply look puzzled when we ask them relative to what they meant to say that X. Just as they would in the case of "Joe is male". If we ask a speaker relative to what they meant to say that Joe is male they will give us at best a strange look. Whether or not the case of hidden relativity can occur will be debated shortly.

Since contextualists about knowledge ascriptions claim that knowledge ascriptions are semantically context sensitive, we can distinguish two parts in such an epistemological theory. The first part might be called the philosophy of language part of the theory and consists in spelling out how it should be understood that knowledge ascriptions are semantically context sensitive. To do this one might either claim that knowledge ascriptions belong to one of the well understood classes of semantic context sensitivity, or describe a new such class. The second part of contextualism about knowledge ascriptions might the called the epistemology proper part of contextualism. In this part contextualists have to spell out what aspects of the context contribute to the content of the knowledge ascription, and how they contribute to the content. In practice contextualists focus much more on the epistemology proper part, and say very little about the philosophy of language part of their theory. In this paper we will look almost exclusively at the philosophy of language part of contextualism. This part brings with it the worry that the kind of semantic context sensitivity that the contextualist claims knowledge ascriptions exhibit is in conflict with certain general truths about language use. We will look at one argument by Stephen Schiffer to that effect in a moment. First we will have to get more clear about how the contextualist wants to use the semantic context sensitivity of knowledge ascriptions to ease the tension between our ordinary claims of knowledge and our inability to answer the skeptic.

2. CONTEXTUALIST SOLUTIONS TO THE SKEPTICAL PARADOX

As I said above, one of the motivations for contextualism is that it seems to have the potential to reconcile our ordinary claims of knowledge with our claims of lack of knowledge in the case of skepticism. But how can this be true? To answer this essentially comes down to answering the skeptical paradox. This paradox consists in a set of three sentences that apparently express propositions that are inconsistent with each other, but that also have quite some plausibility when considered by themselves. Consider:

(I) I know that I have hands.

(II) If I know that I have hands then I know that I am not a brain in a vat.[5]

(III) I don't know that I am not a brain in a vat.

It seems that (I) is just a trivial ordinary claim of knowledge which is true, (III) is a standard claim of lack of knowledge, also true and well motivated by skeptical considerations, but (II) seems to allow one to infer from (I) to the negation of (III). It seems that (I)-(III) are inconsistent. To solve the skeptical paradox, or in Schiffer's terminology[6], to give a happy-face-solution of it, is to show why these really aren't all true or why they really aren't inconsistent. Either one has to show which one of the three isn't true, or why the apparent inconsistency between them isn't real.

This alone, however, is still not enough for a complete solution, as everyone in the debate notices. One also has to give an account of why it seemed to us that we are dealing with an inconsistency here. We have to diagnose what went wrong in our thinking about these matters so that we were fooled into believing that we are dealing with a paradox here. So, to give a happy face solution of the skeptical paradox one has to do both:

(A) Say which one of (I)-(III) is really false, or why they are consistent after all.

(B) Say what fooled us into believing that (I)-(III) are true and inconsistent.

The contextualist solution to the skeptical paradox is the following[7]. Since knowledge ascriptions are context sensitive we have to be careful what the relevant standards of knowledge are that are unarticulated constituents of the propositions expressed. As we saw above, ordinary claims of knowledge are true relative to low standards, and skeptical claims of lack of knowledge are true relative to high standards. So (I) is true relative to low standards, and (III) is true relative to high standards. Now, (II) seems to be true both to high and to low standards. But which ever we take it to be, we will not get an inconsistency. If we read it relative to low standards then it implies together with (I) that we do know relative to low standards that we are not a brain in a vat. That, however, is perfectly consistent with that we do not know this relative to high standards. And if we read (II) relative to high standards then it, together with (III) implies that I do not know relative to high standards that I have hands. This, again, is perfectly consistent with knowing this relative to low standards.[8] This is the contextualist answer to (A).

The contextualist answer to (B) is simply that we are not aware of the context sensitivity ofknowledge ascriptions, and of subtle context changes that occur when we debate skepticism. And because we are not aware ofthis it seems to us that what we say in (I)-(III) is in contradiction. Our ignorance of the context sensitivity and the context changes that occur in the debate about skepticism fool us into thinking that we are dealing with a paradox even though there is no inconsistency here.

3. THE MEANING-INTENTION PROBLEM

Stephen Schiffer argued[9] that the contextualist solution to the skeptical paradox can't work because the two parts ofthis solution, the answer to (A) and the answer to (B), can't both be true. His argument comes down to that according to the contextualist solution knowledge ascriptions are a case ofsemantically context sensitive sentences where the speakers are not aware of the fact that they are semantically context sensitive. This, Schiffer argued, violates general truths about language, namely that what the content of an utterance is has to be backed up with communicative intentions that the speaker has. If the speaker has no communicative intentions to ascribe knowledge relative to low or high standards then the content of the utterance can't be about such standards either. The answer to (B) refutes the answer to (A).

In an earlier paper about the semantics of belief ascriptions Schiffer used a very similar argument and there he entitled it *the meaning-intention problem.*[10] This argument is very plausible for many cases of semantic context sensitivity, and Schiffer goes through a few of them and makes it plausible. For example, he considers the case that was called implicit relativity above, the case where the proposition expressed contains an unarticulated constituent and the speaker is aware that this is so. It seems that in this case the speaker has to have cognitive access to relative to what they meant to say that X. The speaker has to have access to where they meant to say it was raining when they say "It's raining". When we ask them where they meant to say it was raining, we will get an answer. Whatever place the speaker intended to talk about will be the one that is the unarticulated constituent. It could not happen that the speaker is mistaken about what place they were really talking about. In analogy, it could not be that the speaker thinks they are talking about knowledge relative to low standards, whereas they are really talking about knowledge relative to high standards. But something like this seems to be required for the contextualists solution to the skeptical paradox.

Schiffer's conclusion can be nicely stated using the terminology introduced above. According to Schiffer, the proposition expressed by an utterance can only contain a certain unarticulated constituent if the speaker has intentions backing up that this is so and what this constituent is. So, according to Schiffer, the only case ofunarticulated constituents can be the case ofimplicit relativity. According to him, the case ofhidden relativity is impossible. It can not be that an utterance expresses a proposition with an unarticulated constituent, but the speaker would not know what to say to the question relative to what they meant to say that X. In Schiffer's words:

> "For the speaker would not only have to be confounding the proposition she's saying; she'd also have to be totally ignorant ofthe sort of thing she's saying. One who implicitly says that it's raining in London in uttering "It's raining" knows full well what proposition she is asserting; if articulate, she can tell you that what she meant and was implicitly stating was that it was raining in London. But no ordinary person who utters "I know that p", however articulate, would dream of telling you that what he meant and was implicitly stating was that he knew that p relative to such-and-such standard." (Schiffer 1996, p. 326f.)

The situation Schiffer finds unacceptable is exactly the situation of hidden relativity, where the proposition expressed by an utterance of a speaker contains an unarticulated constituent even though the speaker is unaware of that this is so and what this constituent is.

4. THE CASE FOR HIDDEN RELATIVITY

It seems to me, however, that there is good reason to assume that hidden relativity actually occurs. There are examples where we have good reason to assume that they are a case ofhidden relativity. This reason is not conclusive, in the sense that this is the only possible analysis of these examples. But I think it is the best and most reasonable analysis. I will give two such examples. Both examples are based on the same idea, namely that sometimes it is an empirical discovery that a relation that we thought was an n-ary relation, and that we represented as an n-ary relation, really is an n+1-ary relation. This is, in fact, not uncommon. Sometimes it is a simple discovery, and sometimes it is a substantial discovery.

As a first example, consider the case ofmotion. Naively it seems that moving 15 mph is a property of objects, and we usually talk about motion as if this were so. But this is false. Moving 15 mph is a relation between an object and a frame of reference. Thus moving 15 mph is not a unary property, but a binary relation. That this is so is an empirical discovery, namely the discovery that there is no absolute space and no objectively distinguished frame of reference. Thus there is no such thing as absolute motion, and thus there is no such thing as moving 15 mph *simpliciter*. That this is so is pretty much common knowledge these days, but that was not always so. There was a time when people believed that the earth was at rest and that motion is absolute motion. If we would ask a speaker of "I move 15 mph" relative to what frame of reference they meant to say that they move 15 mph we would get as an answer that they did not mean to talk about frames of reference, just about how fast they move.

There are at least four ways to try to deal with this situation. First, we could say that such speakers express the proposition that they have the property of moving absolutely 15 mph and thus the proposition expressed by their utterance contains only two constituents, the speaker and the property of moving absolutely 15 mph.

Secondly, we could say that the speaker only expressed an incomplete proposition, where one of the constituents necessary to complete it was left undetermined.

Thirdly, we could say that these speakers are not talking about motion, but about something else.

Fourthly, we could say that the propositions they express contains the relation of relative motion, the speaker and a frame of reference that is an unarticulated constituent.

It seems to me that the fourth option is the best. The first one seems to make all statements about motion of people who are not aware of the relativity of motion false. Nothing has the property of moving absolutely 15 mph since there is no such thing as absolute motion. In fact, given the laws of physics, there could not be such a thing.

The second option leaves all such statements truthvalueless, since a proposition that is incomplete can't be evaluated with respect to its truth value.

The third one has at least two problems. On the one hand the problem that it doesn't seem right that the people who do not know that motion is relative talk about something else when they use the same word. On the other hand the problem that it seems to be a discovery about motion, and not something else, that it is relative, and not absolute as was generally assumed.

The fourth option has none of these problems. According to it, there will be a frame of reference as an unarticulated constituent in the proposition expressed. The proposition will thus be about relative motion, even though speakers do not know that motion is relative. But which frame of reference is contributed to the proposition as an unarticulated constituent? The following seems to be a reasonable assumption. There is one distinguished frame of reference for the linguistic community using the word "motion". It is the default frame of reference, and this default is the same for everyone, namely the surface of the earth. The important point, I think, is that this frame of reference is the distinguished one for everyone in the linguistic community. It will be the unarticulated constituent for everyone, thereby leading to no weird results about disagreement, or agreement, between speakers. So, when one person utters "Joe is moving 15 mph", and another person utters the same sentence, the propositions expressed will be the same.

Let me give you another example.[11] Being a summer month seems to be a property of months. August has it, December lacks it. But as it turns out, and as was discovered, it is a relation between a month and a hemisphere. August is only a summer month in the Northern hemisphere. It is a winter month in the Southern hemisphere. It seems that before this relativity was discovered people were talking about summer months with no intentions about unarticulated constituents. But within a linguistic community that was entirely in the Northern hemisphere it seems reasonable to say that the speakers there expressed complete propositions with their utterances of "August is a summer month", or "It's summer now". Once the members of a linguistic community entirely located in one hemisphere travel to the other hemisphere they encounter that being a summer month is a relation between a month and a hemisphere.

All this seems to make perfect sense, in particular if we consider that one of the reasons why it was not discovered earlier that being a summer month is a relation is that all the people within a linguistic community lived in the same hemisphere. And if you are always within one hemisphere being a summer month seems to be an absolute property.

It seems to me that the most plausible analysis of these cases is that the propositions expressed by utterances of "I move 15 mph" and "It's summer now" contain unarticulated constituents even though the speakers of these utterances are not aware of the fact that this is so. In other words, these are cases of hidden relativity.

5. THE MEANING-INTENTION PROBLEM, AGAIN

Does the acceptance of the above examples as cases of hidden relativity force one to deny intention based semantics and roughly Gricean ideas of meaning? I don't think so. To see this we have to look at the meaning-intention problem again. This problem is based on the belief that meaning has to be backed up by speaker's communicative intentions. There are two readings of this claim. One of them is in contradiction with hidden relativity, the other one is compatible with it. However, only the second one is well motivated. Let me explain.

Suppose you believe, which seems to be quite reasonable, that the proposition expressed by an utterance of a speaker is determined by the speaker's communicative intentions. Let's call this the thesis of Intention Based Content (IBC). It could be understood in at least two ways. On one reading of it the speaker has to have intentions about what exactly the proposition expressed is supposed to be, and this is understood in the sense that there will be no constituents in the proposition expressed that are not explicitly intended to be part of the proposition by the speaker. Lets call this *the strict reading*. According to the strict reading the speaker will know what the proposition expressed looks like and what its constituents are.[12] This reading of IBC is incompatible with hidden relativity. If hidden relativity occurs then the proposition expressed contains a constituent without the speaker being aware that this constituent is part of the proposition. However, I do not see how such a strict reading of IBC is motivated by the motivation for IBC in general.

There is a much more plausible reading of IBC, *the loose reading*. According to it, it is required for the utterance to have the content that p, that this content is reflected in the right way in the communicative intentions of the speaker. How it is reflected will depend on what the correct theory is of how the content of an utterance is based on communicative intentions. Simply put, an utterance can only have the content that p if the speaker intended to communicate that p with that utterance. The real answer will of course be much more complicated, but this does not matter right now.[13] What is important here is that such a story is perfectly compatible with hidden relativity. The reason is that hidden relativity will apply to the content of mental states just as well as to the content of utterances. When someone ignorant of the relativity of being a summer month believes that it is summer now then the content of this belief will contain an unarticulated constituent, too. In the situation of such a belief we will have the same problem and the same options of solving this problem as we had above in the case of an utterance. And the same reasons will speak in favor of an unarticulated constituent solution to this problem. However, in the case of mental states these constituents should better be called unrepresented constituents. So, it is perfectly consistent with the theory of unarticulated constituents that an utterance can't have the content that p unless the speaker has the intention to communicate that p. The theory of unarticulated constituents applies to mental states as well as utterances. It is in no way required that the person having the intention is aware of all the constituent of the proposition that is, or models, the content of this intention. The loose reading of IBC is well motivated through general consideration in the philosophy of language, but perfectly compatible with hidden relativity.

6. HIDDEN RELATIVITY AND CONTEXTUALISM

We have seen that a contextualist theory about knowledge ascriptions has two parts, a philosophy of language part and an epistemology proper part. In order for contextualism to be able to answer the skeptical problem a contextualist will have to claim that the philosophy of language part involves something like hidden relativity. There has to be an aspect of the content of an utterance of a knowledge ascription that the ordinary speaker is unaware of (in the strict sense spelled out above). Schiffer argued that this can't be right, since it violates general truths about language use. According to him, hidden relativity can't occur. We saw, however, that there are cases that should be understood as cases of hidden relativity. Schiffer's very general worry is thus no refutation of contextualism. However, this should be taken with caution. Schiffer's worry was that hidden relativity can never occur since it violates general truth about language use. We saw that hidden relativity can occur in certain circumstances. The question remains whether or not knowledge ascriptions can be understood as a case of hidden relativity. As a first step towards answering this we should ask whether or not knowledge ascriptions are similar or dissimilar in certain respects to the cases of hidden relativity we saw above. And here one dissimilarity comes to mind right away. In all the cases of hidden relativity we saw above the unarticulated constituent that was contributed to the content without the (strong) awareness of the speaker was the same one for everyone within the language community. Take the case of "summer month". In the example everyone in the language community was within the same hemisphere.[14] And everyone's summer month ascriptions had the same unarticulated constituent in the proposition expressed (the hemisphere they are in). This has the following effect. Whenever two speakers in that language community utter "August is a summer month" they will have produced an utterance with the same content. And whenever one speaker utters "August is a summer month" and another speaker utters "August is not a summer month" the contents of their utterances will be incompatible with each other, in the sense that not both of them can be true. Thus when it comes to sameness and difference (or incompatibility) of contents, the unarticulated constituent doesn't make a substantial contribution. It cancels out, since it is the same for everyone. Not so in the case of knowledge ascriptions, as understood by the contextualist about knowledge ascriptions. According to the contextualist, the contextual contribution to the content of a knowledge ascription will differ substantially form utterance to utterance within the language community. According to the contextualist, it is not so that when one speaker utters "A knows that p" and another speaker utters the same sentence then the content of the two utterances will be the same. And it is not so that if one speaker utters "A knows that p" and another speaker utters "A does not know that p" then the contents of these two utterances are incompatible. Furthermore, according to contextualism, the speakers won't be aware of these facts about difference and compatibility of contents. This follows from the fact that ordinary speakers are not aware of the semantic context sensitivity oftheir knowledge ascriptions, and from the claim that lots of details of the context are relevant for what the content ofa knowledge ascription is.[15]

This is simply an observation of a disanalogy between the cases of hidden relativity we saw above and what a contextualist will have to claim holds of knowledge ascriptions. A contextualist will thus not only hold that speakers have no access (in the strong sense spelled out above) to the content of their utterances, but also no access to sameness, difference and incompatibility ofthe contents of their utterances. This is necessary to run the contextualist solution to the skeptical problem. I think that this, not hidden relativity per se, is the really problematic aspect ofthe philosophy oflanguage part of a contextualist theory about knowledge ascriptions. It is one thing to deny that speakers have access to the content of their utterances in the strong sense spelled out above. After all, it often is an empirical questions what the arity of the relations is that they talk about. It is quite another thing to deny that sameness, difference and incompatibility of contents of utterances is inaccessible to ordinary speakers. But what exactly is wrong with that? And how does it affect contextualism as an epistemological theory?

7. FURTHER ISSUES

We are left with two questions:

1) Is the inaccessibility ofsameness, difference and incompatibility of contents to the speakers in conflict with general truth about language use? And if yes:

2) Does this show that contextualism as an epistemological theory is mistaken, or only that the formulation that contextualism was given, namely as a theory about the content ofknowledge ascriptions, is mistaken?

These are difficult questions that I can't address here in a satisfactory way.[16] However, I can't help but saying what I think is the correct answer to these questions. I think that Schiffer was on the right track in finding the philosophy of language aspect ofcontextualism about knowledge ascriptions rather fishy. In fact, I think that the answer to question 1) is "yes". However, and, I guess, contrary to Schiffer, I don't think this shows much about contextualism as an epistemological theory. I think that the choice to develop the relevant alternatives approach in epistemology as a theory about the content ofknowledge ascriptions was a mistake. However, all that is well motivated about contextualism about knowledge ascriptions does not rely on contextualism having this formulation. The philosophy oflanguage ofcontextualism about knowledge ascriptions is wrong, but this doesn't affect the epistemology of it, or at least not in a bad way. I will spell this out in Hofweber (n.d.).

T. Hofweber
Stanford University
U.S.A.

NOTES

* Thanks to Keith DeRose, Fred Dretske, John Perry and Stephen Schiffer for discussion and comments, and to the organizers of the 5th International Colloquium in Cognitive Science (ICCS-97) to invite me to give a talk on contextualism.

[1] See Dretske 1970, 1971, 1981, Goldman 1976.
[2] See Cohen 1991.
[3] For example Cohen 1988, DeRose 1992 and Lewis 1996.
[4] See Perry 1986.
[5] This second premise sometimes gets formulated as
(II*) I know that if I have hands then I am not a brain in a vat.
This is a subtle but potentially important difference. It does not matter for the present discussion, however, and I will ignore it.
[6] See Schiffer 1996.
[7] See DeRose 1995 for a much more detailed solution like the one sketched here. DeRose does not use the terminology of unarticulated constituents. He could have, though, and it is perfectly consistent with his theory to do so.
[8] I ignore the issues about whether or not considering brains in a vat automatically raises the standards to high standards. This is not relevant for the present discussion, even though a central issue in contextualism and how skepticism works.
[9] In Schiffer 1996.
[10] See Schiffer 1992.
[11] I owe it to John Perry.
[12] At least if we assume that the speakers have cognitive access to their communicative intentions, which seems reasonable in this context.
[13] For Grice's account of this see Grice 1989, and Schiffer 1972.
[14] Of course, the example was a simple case, and in real live with global travel and the like, language communities go across hemispheres. This makes the situation more complicated, in particular what we should say what the content of an summer month ascription is in, say, a phone call across hemispheres. The simple example above was simply one where it seems plausible that the is a case of hidden relativity.
[15] See DeRose 1992, for example. Contextual features that are supposed to be relevant are, for example, importance to know, which is not easily accessible to the hearer.
[16] They are addressed in detail in my "Contextualist Theories of Knowledge and of Knowledge Ascriptions", in preparation.

REFERENCES

Cohen, Stewart. 1988. "How to be a Falibilist". *Philosophical Perspectives* 2:91-123.
Cohen, Stewart. 1991. "Skepticism, Relevance and Relativity" in *Dretske and his Critics* Brian McLaughlin (ed.). Basil Blackwell.

DeRose, Keith. 1992. "Contextualism and Knowledge Attributions". *Philosophy and Phenomenological Research* ***52***:913-29.

DeRose, Keith. 1995. "Solving the Skeptical Problem" *The Philosophical Review* ***104***: 1-52.

Dretske, Fred. 1970. "Epistemic Operators" *Journal of Philosophy* ***67***: 1007-23.

Dretske, Fred. 1971. "Conclusive Reasons" *Australasian Journal of Philosophy* ***49***: 1-22.

Dretske, Fred. 1981. "The Pragmatic Dimension of Knowledge" *Philosophical Studies* ***40***: 363-78.

Goldman, Alvin. 1976. "Discrimination and Perceptual Knowledge" *Journal of Philosophy* ***78***: 771-91.

Grice, Paul. 1989. *Studies in the Way of Words* Harvard University Press.

Hofweber, Thomas. n.d. "Contextualist Theories of Knowledge and of Knowledge Ascriptions". In preparation.

Lewis, David. 1996. "Elusive Knowledge". *Australasian Journal of Philosophy* ***74***: 549-567.

Perry, John. 1986. "Thought without Representation". Reprinted in Perry 1993.

Perry, John. 1993. *The Problem of the Essential Indexical and Other Essays,* Oxford UP.

Schiffer, Stephen. 1972. *Meaning* Oxford: Clarendon Press.

Schiffer, Stephen. 1992. "Belief Ascription" *Journal of Philosophy.* ***LXXXIX, 10***: 499-521.

Schiffer, Stephen. 1996. "Contextualist Solutions to Skepticism". *Proceedings of the Aristotelian Society* ***XCVI:*** 317-333.

FERNANDO GARCÍA MURGA

PRESUPPOSITIONAL AND RHETORICAL EFFECTS OF FOCUS PARTICLES: THE CASE OF 'EVEN'.

Abstract. It is the objective of this paper to present a general account of the semantics and pragmatics of the word 'even'. As a general assumption, the word 'even' is lumped together into the set of 'focal particles', i.e., lexical items whose main function consists of 'highlighting' a linguistic expression. The main theses proposed in this paper are the following: First, 'even', being a focal particle, triggers a focal presupposition, which is equal to the focal presuppositions derived from stress and cleft sentences, except for the quantificational force specific to 'even'. This focal presupposition will be dubbed 'the plurality condition'.

Second, 'even' displays an argumentative force which serves the speaker to exploit general background knowledge (comparable in part to Aristotlean topoi) to guide the addressee toward a specific conclusion. It is claimed here that the 'violation of an expectation condition' that appears in classical analyses of 'even' is just a part of the argument that the addressee has to build. In some sense, the background knowledge *recalled* by the word 'even' to build up an argument is also presuppositional. Therefore, the presuppositional elements attached to an 'even' containing sentence go beyond what has been assumed up to now.

Finally, an important result of this paper is the 'redemption' of the term 'presupposition' from the wastepaper basket where it was thrown in the late seventies.

1. INTRODUCTION

It has been said that even words give trouble, even to linguists! In this paper, then, an analysis of the semantics and pragmatics (if these realms may be differentiated) of the word 'even' is proposed.

One of the initial assumptions is that the word 'even' is a member of the set of 'focal particles', i.e., lexical items whose main function consists of specifying the sentence focus. Unfortunately, the notion of focus has received very different characterisations.[1] The most pervasive characterisation is based on informative features; that is, it is assumed that the focus represents the 'new information' provided by the utterance. The position I advocate in this paper is that focus creates a focus/presupposition partition of an utterance. However, this thesis will be maintained as far as a reanalysis of the terms 'focus' and 'presupposition' is carried out.[2]

Conceivably, it is expected then, that the semantic and pragmatic analysis of 'even' will coincide in the aspects related to 'focus' with other focal particles like

K. Korta et al. (eds.), Cognition, Agency and Rationality, 105-121.

'only', 'also', 'too', etc. Moreover, if this paper is moving in the right direction, it should be possible to sketch a unified analysis ofthe large set of focal constructions: (contrastive) stress, cleft sentences, some morphological markers and focal particles.[3]

This paper will initially be centred on the following simple examples (taken from Horn 1969):

(1) Even Muriel voted for Hubert.

In sentence (1), 'even' unambiguously focuses on the NP 'Muriel'. It seems intuitively clear that sentence (1) is somehow related to sentences (2)-(4) below:[4]

(2) Muriel voted for Hubert.

(3) Someone other than Muriel voted for Hubert.

(4) The speaker would not expect or would not expect the hearer to expect Muriel to vote for Hubert.

At a first glance, sentence (2) displays the informative content of (1). That is, (2) seems to be explicitly conveyed by (1). In their turn, sentences (3) and (4) are implicitly communicated by (1), and, in some sense, they are the condition for the appropriateness of the utterance of sentence (1). In order to distinguish (3) and (4), sentence (3) will be dubbed 'the plurality condition', and sentence (4), 'the violation of an expectation condition'.

The first main thesis held in this paper is, thus, that 'even', being a focal particle, triggers a focal presupposition. This contention deserves some comments. In sentence (1), 'Muriel' is focus and, therefore, it is taken as 'new'. But, simultaneously, since 'Muriel' is a proper name, it is taken as 'old'. So, it makes no sense to say that a constituent is simultaneously new and old. This contradictory situation has posed serious problems to the focus/presupposition approach (Schmerling 1976, Tancredi 1992). Nevertheless, it is the identification of the focus as the element that takes part in the situation described which is new (pace further determination ofthis epistemic feature).

Then, if (3), the plurality condition, is a focal presupposition triggered by (1), 'even' is expected to share some properties with the focal presuppositions derived from stress and cleft sentences except for the quantificational force specific of 'even'.

Secondly, it will be contended here that 'even' displays an argumentative force which serves the speaker to exploit general background knowledge (comparable in part to Aristotlean topoi) to guide the addressee toward a specific conclusion. In some sense, the background knowledge *recalled* by the word 'even' to build up an argument is also presuppositional. Now, this second kind of presupposition will be dubbed 'argumentative presupposition'. In the example above, sentence (4), the 'violation of expectation condition' is an argumentative presupposition licensed by (1).

The analysis of 'even' leads us then, to one of those weasel words one cannot use without some trepidation. It is, however, the main aim of this paper to rescue the term 'presupposition' from the wastepaper basket where it was thrown in the late seventies. In trying to do that, the presuppositional framework developed in previous works (García Murga 1995a, 1995b) will be adopted. Basically, in those works, efforts were directed toward the forecast of a unified theory of linguistic presuppositions.

The analysis of the word 'even' offered here makes use of the theoretical tools defined for the analysis of lexical and existential presuppositions in García Murga 1995b. This means that, if the arguments ofthis paper are too not far from the mark, it should be possible to foresee a unified theory not only offocal presuppositions but offocal, lexical and existential presuppositions (i.e. linguistic presuppositions).

At the same time, the analysis of 'even' offers the possibility of drawing a first sharp and grounded distinction between linguistic presuppositions and other kinds of presuppositions related to the rhetoric component of language (Ducrot 1972, Anscombre and Ducrot 1983): the argumentative presuppositions.

This paper will proceed as follows. In the next section, I take up the interpretation of 'even'. Then, three different analysis of 'even' will be (very sketchily) presented and criticised: the classical presuppositional analysis (Horn 1969), the implicature analysis (Karttunen & Peters 1979) and the pragmatic entailment analysis (Fauconnier 1975, Kay 1990).

Then, the following two sections will be devoted to the derivation of the two (allegedly) presuppositional conditions attached to the word 'even'. In section 3, the mechanisms necessary for the derivation of the focal presupposition (i.e., the 'plurality condition' attached to 'even' will be introduced. In section 4, the analysis ofthe argumentative force of 'even' will be taken as the basis for the derivation of the 'violation ofexpectation condition'.

Finally, some conclusions related to presuppositional phenomena and the realm ofSemantics and Pragmatics will be drawn.

2. THE CONTRIBUTION OF 'EVEN' TO SENTENCE MEANING

The example discussed in the first section had the structure '[Even NP] [VP]'. In all these cases, the particle 'even' invariably focuses on the NP, so that the plurality condition comes unambiguously from the substitution of the focused NP by an appropriate variable.[5] However, 'even' may appear—amongst other syntactic positions—after the NP subject:

(5) Muriel even voted for Hubert.

Putting to one side the interpretations of'even' as focusing the subject in (5) (an interpretation that require the prosodic prominence lying on 'Muriel'), sentence (5) is two-ways ambiguous with respect to the scope of 'even'. Examples (6) and (7) below illustrate the two relevant readings of (5). In (6), 'even' focuses on the NP 'Hubert'. In (7), the whole VP 'voted for Hubert' is the focus of'even':

(6) Muriel used to vote for her friends. This time, Muriel even voted for Hubert.

(7) Muriel used to campaign for Hubert. This time, Muriel even voted for Hubert.

In this section, not all the possible data related to 'even' has been reviewed. A noteworthy absence is the 'even if' concessive conditional structure which, obviously, deserves deep analysis of its own. Now, attention will be paid to the theoretical mechanisms that derive (3) and (4) from (1) above.

2.1. The classical presuppositional analysis.

According to Horn (Horn, 1969), sentence (1) asserts (2) and presupposes (3) (repeated here for convenience):

(1) Even Muriel voted for Hubert.

(2) Muriel voted for Hubert.

(3) Someone other than Muriel voted for Hubert.

Besides this, sentence (1) warrants the sentence (4) below (adapting Fraser's formulation in Fraser (1970); repeated here for convenience):

(4) The speaker would not expect or would not expect the hearer to expect Muriel to vote for Hubert.

It is worth remembering that 'presupposition' used to be defined as the following logical relation:

(def. 1) A presupposes B if and only if (i) A entails B and (ii) ¬A entails B.

In its turn, entailment may be defined as follows:

(def. 2) A entails B if and only if whenever A is true, B is true.

Even though Horn's presuppositional analysis is intuitively appealing, the problem ofthe relation between sentence (1) and sentences (3)-(4) arises early on. Note that, if sentences (3) and (4) are taken as presuppositions triggered by sentence (1) then it is predicted to be a logical relation ofentailment between (1) and (3)-(4). Hence, from the definition (1) above, it follows that if sentence (1) is true then sentences (3) and (4) are true too.

However, Kempson (1975) showed that sentences (3) and (4) cannot be logically worked out from (1) because (1) may be true even if (3) or (4) are false. A case at hand is when nobody but Muriel voted for Hubert. In such a case, sentence (2), which is the asserted part of(1), is true. Now, if it is assumed that 'Even A is P' is true wherever 'A is P' is true, then (1) is true under the context that is being considered. At the same time, (3) is obviously false.

Similarly, it is easy to see how the context may be altered so that (1) remains true but (4) becomes false. It may still be contended that sentence (1) is inappropriate when (3) and (4) are false. But it has been argued that the following example (adapted from Kempson 1975) is correct even though there are no expectations about Max trying on fancy pants and nobody else tried on fancy pants:

(8) All the kids tried on something. Mary tried on a fancy tie. Sue a long shawl. Even Max tried on fancy pants. Now that his mother has married again, it's not really surprising because he joins in with things much more.

The specificity of the sentence containing 'even' in (8) lies in the fact that it does not imply that 'a plurality of persons tried on fancy pants'. Certainly, it may be argued that, when the sentence containing 'even' is uttered, the context contains the information 'Somebody tried on something'. Then, it may be thought that the fancy pants Max tried on are an instantiation of the variable 'something'. So, the argument goes, all what we need here is a more explicit theory of discourse.

However, it is still true that the sentence containing 'even' in (8) has still another interpretation if uttered in a different context. Being tentative, 'Even Max tried on fancy pants' may be understood as follows:

(9) It is Max's type of strange behaviour that he, being as he is, tried on fancy pants.

This interpretation leaves open the possibility that 'nobody else tried on fancy pants'. In sum, since according to classical theories of presuppositions, presuppositions are the result of a logical relation, it is concluded that neither (3) nor (4) are presuppositions of sentence (1).

A second main argument against Horn's presuppositional analysis came from negative sentences containing 'even'. Consider the following sentence:

(10) (It is) Not (the case that) even Muriel voted for Hubert.

which, according to Horn, is synonymous to (11) below:

(11) Even Muriel did not vote for Hubert.

Now, sentences (10) and (11) are interpreted not only as negating (2), but also as negating (3) and (4). That is, the second condition for presuppositional relation to hold in the definition (def. 1) above is not fulfilled. In other words, if, as all classical theories of presupposition hold, negation does not affect presupposition under the preferred reading, (3) and (4) cannot be presuppositions triggered by (1).

Horn's way out was to disregard (10) and (11) as the negation of (1). This is, to my mind, a counter-intuitive move. Moreover, Horn's solution led him to accept the striking point that it is impossible to negate the assertion of 'even' containing sentences and to affirm their presuppositions simultaneously.

To sum up. Horn's presuppositional analysis of 'even' is not compatible with the classical (logical) definition of presuppositions. In fact, examples were found such that neither the condition (i) of the definition (def. 1) nor the condition (ii) were fulfilled.

This fact contributed to the abandonment of the notion of 'presupposition' in analysing 'even'. However, all that has been said here militates against the analysis of 'even' based on the logical relation of presupposition as defined by the definition (def. 1) above. But, since the logical definition of presupposition had to be abandoned for independent reasons,[6] the examples adduced here cannot be taken to show that sentence (1) does not presuppose (3)-(4).

Now, once it was thought that (3) and (4) were not truth-conditionally related to (1), it was claimed that (3) and (4) were non-truth conditional aspects of the meaning of (1).

2.2. The implicature analysis.

One of the main results of Gricean theory of conversation is that there are a large set of non-truth conditional meaning relations in natural languages. Since the previous section has shown that the contribution of the word 'even' to the meaning of the sentence cannot be truth conditional, 'even' becomes a good candidate for licensing a non truth-conditional relation.

The first steps in this direction were taken by Karttunen and Peters (1979). According to them, the plurality condition and the violation of an expectation are systematically attached to the word 'even', i.e., (3) and (4) are conventional implicatures of (1).[7] However, they are derived following different paths.

The plurality condition (3) comes from the focal nature of the word 'even'. Informally stated, the word 'even' identifies a particular focus. Then, condition (3) obtains by deleting the focus and filling the hole with an appropriate variable.

The violation of expectation condition (4) is produced via the formulation of an scale, which is just an ordered set of linguistic expressions. Karttunen and Peters define the scale relevant to 'even' as follows:

> "For all x under consideration besides a, the likelihood that ...x... is greater than the likelihood that ...a..." (Karttunen & Peters 1979: 26)

Then, if this conventional implicature[8] is applied to example (1), it follows that 'For all people that voted for Hubert, the likelihood that each one of those people voted for Hubert is greater than the likelihood that Muriel voted for Hubert'. That is, Karttunen and Peters maintain that the focus marked by 'even' points out the lower element in the relevant scale. Therefore, in example (1), Muriel is the least likely person to vote for Hubert.

Karttunen and Peters' implicature theory, then, predicts that the focus marked by the word 'even' is taken by the speaker as the least likely thing that fulfils the corresponding predicate. However, the following example contradicts such a prediction:

(12) Ronald possesses a Jag, a large flat in Manhattan and even a private jet.

In sentence (12), the fact that Ronald has a Jag and a large flat in Manhattan indicates that Ronald is very rich. Then, a private jet is not the least likely thing that a wealthy person may afford. Note, moreover, that it is possible to alternate the conjoined elements in (12):

(13) Ronald possesses a private jet, a Jag, and even a large flat in Manhattan.

Another point that merits some comment is the role that sentences (3) and (4) play in the acceptability of sentence (1) in a context. Karttunen and Peters stress that the plurality condition and the violation of expectation condition are not conditions of truth, but conditions of use; i.e., (3) and (4) are felicity conditions for the utterance of (1). Moreover, as Karttunen and Peters say, (3) and (4) should be part of the common ground that the speaker and the addressee share when (1) is uttered. That is, (3) and (4) have the crucial features of the 'pragmatic presuppositions':

> "Surface sentence A pragmatically presupposes a logical form L, if and only if it is the case that A can be felicitously uttered only in contexts which entail L" (Karttunen 1974: 149)

> "A proposition *P* is a pragmatic presupposition of a speaker in a given context just in case the speaker assumes or believes that *P*, assumes or believes that his addressee assumes or believes that *P*, and assumes or believes that his addressee recognises that he is making these assumptions, or has these beliefs" (Stalnaker 1974: 137)

Karttunen and Peters argue that 'pragmatic presupposition' is not a primitive notion, but a derived one. In particular, they claim that the (pragmatic) presuppositional behaviour of (3) and (4) may be explained with the tools offered by the Gricean theory of conversation. Specifically, Karttunen and Peters appeal to the cooperative principle[9] and to the following maxim:

> "A sentence ought to be uttered only if it does not conventionally implicate anything that is subject to controversy at that point in the conversation" (Karttunen & Peters 1979: 14)

It has to be conceded that the notion 'pragmatic presupposition' is far from being a well-defined notion. First, the substitution of the idea of presuppositions as 'conditions of truth' by the expression 'conditions of use' does not resolve the problems mentioned in the note 6 above (see García Murga 1995a, chapter 3). Besides, Stalnaker's definition cannot avoid the conundrum of 'mutual knowledge'.

However, the theoretical foundation of the notion of 'implicature' is weak in as much as it constantly requires ad hoc principles or maxims. The following section deals with that, and even though it may be taken as an extension of the implicature theory, makes use of a very attractive notion: 'pragmatic entailment'.

2.3. The pragmatic entailment analysis.

As has been shown in the previous section, it is widely assumed that the word 'even' is interpreted as carrying a scale and pointing out some low element in the scale. This section will offer a more fine-grained analysis of the notion of 'scale'.

Scales are built up pragmatically. So, going back to example (1), the speaker, by uttering (1), bears in mind a set of people that voted for Hubert, say S1 {Muriel, Susan, Peter, John}. Now, according to known facts about those persons, the speaker creates a scale ranging from those who are expected to vote for Hubert (for instance, friends, relatives, etc.[10]) to those who are not expected to vote for Hubert (for example, enemies, etc.). The following scale, then, orders the set S1 according to the relation R ('to vote for Hubert'):

Peter
Susan
John
low point: Muriel

Now, according to Fauconnier (Fauconnier 1975: 364), the following rule may be stated:

Rule 1: If x1 is lower than x2 then R(x1) entails R(x2)

Therefore, to build up pragmatic scales, S1 has to contain more than one element. The plurality condition, thus, follows directly from the pragmatic scale, as may be easily observed in example (1).

However, as was indicated in the previous section, the end-of-scale condition is excessively strong. Kay (1990) has modified this end-of-scale condition, but has retained the notion of 'scale' and the notion of 'entailment'. The basic tools of Kay's theory are the following:

(i) the sentence that contains 'even' (tp) expresses a proposition that is more informative than some other proposition contained in the context (cp).

(ii) p is more informative than q if p entails q in a given scalar model.

(iii) a scalar model is, loosely speaking, a formalization of Fauconnier's scales.

Now, Kay's theory predicts that 'even' can appear only in sentences in which the tp entails the cp, given a scalar model. However, sentences (12) and (13) are obvious counterexamples because for example, to have a private jet does not entail having a Jag and a large flat in Manhattan in a normal scalar model for possessed things. Kay tries to overcome this flaw by postulating a 'totalization process'. According to this 'totalization process', the cp and the tp relevant for the entailment relation are 'the sum' of the things that pertain to the scale. To illustrate, the cp and the tp of sentence (12) will be, respectively:

(12') a. Ronald possesses a Jag and a large flat in Manhattan, i.e., Ronald possesses two things. (cp)

b. Ronald possesses a Jag, a large flat in Manhattan and a private jet, i.e., Ronald possesses three things. (tp)

Now, under this analysis, the tp entails the cp in sentence (12). However, this proposal goes too far since the totalization process leaves the theory vacuous because the totalization process makes licit any sentence containing 'even'. Note, for instance, that the following sentences cannot be distinguished (Anscombre & Ducrot (1983)):

(14) Marie knows lots of languages: English, German, Spanish, and even Hebrew.

(15) ?? Marie knows lots of languages: Hebrew, German, Spanish, and even English.

Simultaneously, Kay's theory does not go far enough because it cannot explain why sentences like (16) below are odd even though there is an entailment relation between the tp and the cp:

(16) ?? She is a basketball player, a tennis player and even a sportswoman.

Finally, even logically contradictory sentences may be joint by 'even' (Anscombre & Ducrot (1983)):

(17) The belligerency ofthe Proletariat has not changed, it has even increased.

Therefore, it is concluded that the pragmatic entailment analysis of 'even' is insufficient.[11] Thus, the remainder of this paper is devoted to developing a new analysis of 'even'. As a first step, the plurality condition will be examined in the next section. Then, in section 4, the violation of expectation condition will be analysed in the light ofAnscombre and Ducrot's argumentative theory.

3. 'EVEN' AS A (FOCAL) PRESUPPOSITIONAL CONSTRUCTION.

The first task to be faced in this section is the development of a system of representation capable of handling the data mentioned in sections 1 and 2. As a working assumption, the basic insights of Jackendoffean Conceptual Semantics are adopted.[12] Specifically, what is going to be said here crucially hangs on the decompositionality oflexical items.[13]

For the sake of simplicity, the complexities of the representation of certain expressions, as well as the tense, will be ignored. Thus, the following representation is proposed for sentence (2):

(18)

$$\begin{bmatrix}\text{EVENT}\\ \text{CAUSE}\end{bmatrix}\left(\begin{bmatrix}\text{THING}\\ \text{MURIEL}\end{bmatrix}, \begin{bmatrix}\text{EVENT}\\ \text{HAVE}\end{bmatrix}\left(\begin{bmatrix}\text{THING}\\ \text{HUBERT}\end{bmatrix}, \begin{bmatrix}\text{THING}\\ \text{VOTE}\end{bmatrix}\right)\right)$$

Basically, representation (18) says that Muriel caused Hubert to have a vote. As it has been assumed, 'even' is a focal particle, that is, 'even' is a lexical item that permits the speaker to focus on a specific part of a sentence. In a previous work on the focal properties of stress (García Murga, manuscript), I introduced a marker of focus 'F'. This focal marker becomes attached to the focused conceptual constituent in the interpretation process.

The function of the marker 'F' is twofold. First, 'F' indicates that the implication rule R2 below has to be applied and, second, it forces the addressee to search for the result in his/her activated memory.

Rule 2 (R2)

$$\begin{bmatrix} [\text{Z1}]\ldots & \begin{bmatrix} \text{X} \\ \text{Y} \end{bmatrix} \ldots [\text{Zn}] \end{bmatrix} \rightarrow [[\text{Z1}]\ldots[\text{X}]\ldots[\text{Zn}]]$$

Besides its focal nature, 'even' is an additive particle (König, 1991). This means that 'even' licenses the existence of other values that satisfy the open sentence worked out by means of rule 2. The additive nature of 'even' is specific of a subset of focal particles. However, other very different cases that indicate multiplicity are the morphological plural and the 'the NP + relative clause' structure:

(19) *The politicians* shake thousands of hands when they are campaigning.

(20) *The man that is shaking hands* is a politician.

In sentence (19), more than one politician is mentioned. In sentence (20), the whole NP 'the man that is shaking hands' presupposes that there is more than one man in the environment of sentence performance. This means that the conceptual elements that are introduced below are not 'ad hoc' or specific elements created for the analysis of 'even'.

Now, it sounds reasonable to assume that the representation of 'even' has something in common with the representation of the morphological plural (and of the structure 'the NP + relative clause' as well). Then, adopting Jackendoff's (1996) representation of morphological plural, where '||' represents a projection of a 'representative slice' of a category onto a discrete axis represented by '[PLUR]', the following representation of (1) is proposed:

(21)

$$\begin{bmatrix} \text{EVENT} \\ \text{CAUSE}\left(\begin{bmatrix} [\text{PLUR}] \\ || \\ \text{THING} \\ \text{MURIEL} \end{bmatrix}^{F}, \begin{bmatrix} \text{EVENT} \\ \text{HAVE}\left(\begin{bmatrix} \text{THING} \\ \text{HUBERT} \end{bmatrix}, \begin{bmatrix} \text{THING} \\ \text{VOTE} \end{bmatrix} \right) \end{bmatrix} \right) \end{bmatrix}$$

Now,[14] it is easy to see that the application of rule 2 on (21) produces a representation (22) that can be read as (23) below:

(22)

$$\begin{bmatrix} \text{EVENT} \\ \text{CAUSE}\left(\begin{bmatrix} [\text{PLUR}] \\ || \\ \text{THING} \end{bmatrix}, \begin{bmatrix} \text{EVENT} \\ \text{HAVE}\left(\begin{bmatrix} \text{THING} \\ \text{HUBERT} \end{bmatrix}, \begin{bmatrix} \text{THING} \\ \text{VOTE} \end{bmatrix} \right) \end{bmatrix} \right) \end{bmatrix}$$

(23) A plurality of things (persons) voted for Hubert.

I claim that (23) is a more accurate formulation than (3) of the (alleged) focal presupposition triggered by (1).

Now, this analysis can be easily extended to cover the cases of scope ambiguity. As was noted in the preceding section, sentence (5) has two (relevant) readings, represented, respectively by (6) and (7) (repeated here for convenience):

(5) Muriel even voted for Hubert.

(6) Muriel used to vote for her friends. This time, Muriel even voted for Hubert.

(7) Muriel used to campaign for Hubert. This time, Muriel even voted for Hubert.

The reading biased by the context in (6) corresponds to the following representation:

(24)

$$\begin{bmatrix}\text{EVENT} \\ \text{CAUSE}\left(\begin{bmatrix}\text{THING}\\ \text{MURIEL}\end{bmatrix}, \begin{bmatrix}\text{EVENT}\\ \text{HAVE}\left(\begin{bmatrix}\text{[PLUR]}\\ \| \\ \text{THING}\\ \text{HUBERT}\end{bmatrix}^{F}, \begin{bmatrix}\text{THING}\\ \text{VOTE}\end{bmatrix}\right)\end{bmatrix}\right)\end{bmatrix}$$

In turn, the interpretation of (5) in context (7) comes from the following representation:

(25)

$$\begin{bmatrix}\text{EVENT} \\ \text{CAUSE}\left(\begin{bmatrix}\text{THING}\\ \text{MURIEL}\end{bmatrix}, \begin{bmatrix}\text{[PLUR]}\\ \| \\ \text{SITUATION}\\ \text{HAVE}\left(\begin{bmatrix}\text{THING}\\ \text{HUBERT}\end{bmatrix}, \begin{bmatrix}\text{THING}\\ \text{VOTE}\end{bmatrix}\right)\end{bmatrix}^{F}\right)\end{bmatrix}$$

The application of R2 on (24) and (25) produces respectively the following results:

(26)

$$\begin{bmatrix}\text{EVENT} \\ \text{CAUSE}\left(\begin{bmatrix}\text{THING}\\ \text{MURIEL}\end{bmatrix}, \begin{bmatrix}\text{EVENT}\\ \text{HAVE}\left(\begin{bmatrix}\text{[PLUR]}\\ \| \\ \text{THING}\end{bmatrix}, \begin{bmatrix}\text{THING}\\ \text{VOTE}\end{bmatrix}\right)\end{bmatrix}\right)\end{bmatrix}$$

(27)

$$\begin{bmatrix}\text{EVENT} \\ \text{CAUSE}\left(\begin{bmatrix}\text{THING}\\ \text{MURIEL}\end{bmatrix}, \begin{bmatrix}\text{[PLUR]}\\ \| \\ \text{SITUATION}\end{bmatrix}\right)\end{bmatrix}$$

that can be read respectively as:

(28) Muriel voted for a plurality of things/persons.

(29) Muriel did a plurality of things.

Then, the final interpretation of (5) depends on the context of utterance. In fact, the resolution of the ambiguity crucially hangs on the procedural mechanism triggered by the focus marker 'F' (for more details, see García Murga, manuscript). What is crucial here is that the logical implications of a sentence are not worked out in an 'out-of-the-blue context'. That is, logical implications come from a conceptual structure that is built up -amongst other things- thanks to contextual information. It is only in this sense that it is possible to say that rule 2 represents a 'pragmatic entailment'.

Now, I claim that this basic assumption of what may be dubbed 'Conceptual Logic' serves to overcome the objections against the presuppositional analysis of 'even'. So, the sentence containing 'even' in (8) does not militate against the analysis proposed here. It has to be noted that, in context (8), the relevant point is the trying on of something. So, the NP 'fancy pants' is just an instantiation of the discursive relevant variable (a complete explanation will require a whole discursive theory, which is beyond the aim of this paper).

Again, the interpretation given in (9) does not force one to give up the analysis held here. From (9) a plurality of actions performed by Max follows, so that they define his character. It is, therefore, just a more fine-grained analysis of NP that is required (again, a task beyond this paper).

Finally, it is misleading to utter (1) in a context that does not block interpretation (21) and, simultaneously, consider (22) false (see García Murga, 1995a for a discussion of the 'presupposition failure' and the pragmatics of the notion of truth).

With respect to the negative sentences (10) and (11), their presuppositional behaviour follows directly from the mechanisms that guide the interpretation of negative sentences. In short, (10) and (11) are represented as (30) bellow:

(30) NOT

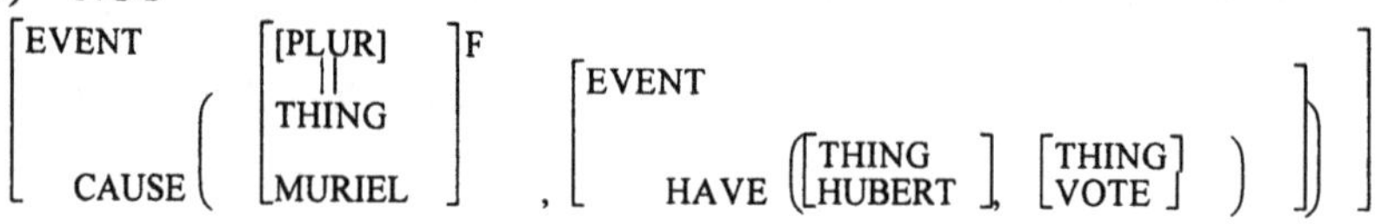

Let us assume the following rule of negation attachment:

Rule 3 (R3)

$$\overset{\text{NOT}}{[\ [X1]\ \ldots\ [Xi]\ \ldots\ [Xn]\ ]} \rightarrow \overset{\text{NOT}}{[\ \ldots\ [Xj]\ \ldots\ ]}$$

The application of rule 3 is purely pragmatic, that is, there is no a priori possibility to know what conceptual constituent will be affected by negation. However, let us assume that the application of R3 on (30) gives rise to (31):

(31)

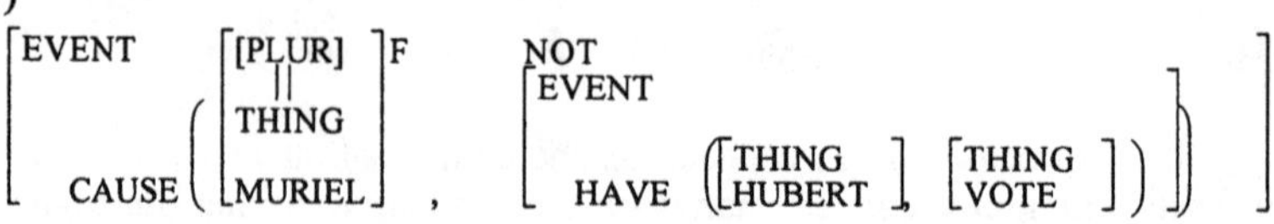

Now, (32) and (33) below are drawn from the application of the rule 2 (R2) on (31):

(32)

```
[EVENT                              NOT
                                    [EVENT
       (                                       ([THING    ] [THING ])  )]
  CAUSE ( [MURIEL],                  HAVE     ([HUBERT   ], [VOTE  ])  )]
```

(33)

```
[EVENT                              NOT
                                    [EVENT
          ([[PLUR]   ]
          (   ||     ]                         ([THING    ] [THING ])  )]
  CAUSE   ([THING    ],              HAVE     ([HUBERT   ], [VOTE  ])  )]
```

These structures mean, respectively:

(34) Muriel did not vote for Hubert.

(35) Someone other than Muriel did not vote for Hubert.

In fact, this is the natural interpretation of the negative sentences (10) and (11). Note that, as Horn observed, it is impossible to deny the assertive part of an 'even' containing sentence and, simultaneously, to maintain the focal presupposition. Now, this observation is easily explained. Note that the assertion and the focal presupposition share the second argument of the function 'CAUSE'. Therefore, according to the representation system adopted here, it is not possible to draw from the same conceptual structure (34) and (3).

In a nutshell, 'even' is a focal presuppositional construction whose behaviour is similar to other focal presuppositional constructions such as (contrastive) stress. Thus, in this section, a representational system and a rule of implication have been defined. Now, the rhetoric effects of 'even' have to be considered.

4. THE RHETORICS OF 'EVEN'.

The previous section has been devoted to the analysis of the plurality condition. It is time now to take up what has been called the violation of expectation condition. In line with Anscombre and Ducrot's argumentative theory (Anscombre & Ducrot, 1983), I maintain that 'even' uncovers an argumentative structure attached to language that is built up from encyclopaedic knowledge. The argumentative structure is used by the speaker to guide the addressee towards some specific conclusion.

In example (1), the range of possible conclusions is hard to constrain. Just as an example, let us suppose that Muriel is the wife of Hubert's opponent in the election (let me call him Johnson). In addition, suppose that what is choosen in the election is the President of the United States. Then, given the normal general knowledge of family relations and of democratic elections, and given a set of parameters necessary for the appraisal of a situation (see Sánchez de Zavala, 1991), sentence (2) violates normal expectations. Sketchily, the reasoning may run as follows:

(36) If Muriel is married to Johnson, Muriel loves Johnson. Then, all situations positively appraised by Muriel in which Johnson is involved, have to be launched by her. The winning of the election for the President of the United States is positively evaluated by Muriel. To win a democratic election, it is necessary to obtain as many votes as possible. Therefore, Muriel is expected to vote for Johnson.

Then, sentence (1) warrants the violation of an expectation, as sentence (4) states. Moreover, the fact that Muriel voted for Hubert may be used by the speaker as a strong argument for a certain conclusion. In the situation described above, the conclusion may be (amongst others) that Johnson is a very bad politician.

Finally, it is worth noting that the interpretation of(1) changes completely if the context changes (i.e., if what is elected is who will take part in a dangerous mission, if Muriel is involved in a process ofdivorce, etc.) or if the cultural and the person's moral norms change. This great context dependence of the interpretation of 'even' containing sentences explains why sentences like (15) and (16) sound odd. Assuming that, nowadays, English is the most extended language in the world and that basketball and tennis are sports, it is difficult to see how the conclusion that the addressee should infer looks like.

What is crucial in the example we began with is that (4) and the premises of the reasoning in (36) can be taken as presuppositions attached to (1) (at least in a non-technical sense ofthe term 'presupposition'). However, they differ considerably from the focal presupposition worked out from (1). Therefore, I claim that a sharp line has to be drawn between 'focal presuppositions' and 'argumentative presuppositions' (the psychological basis ofthis distinction has yet to be investigated).

5. CONCLUSIONS.

In this paper, a presuppositional analysis of'even' has been proposed. Even though it has been sketchily shown in this paper, the presuppositional analysis has required the formulation ofa new presuppositional theory. As the interpretation of sentences containing 'even' reveals, two, very different types of presuppositional relations have to be differentiated: 'linguistic presuppositions' (which include the focal presuppositions) and 'argumentative presuppositions'.

As a provisional definition, linguistic presuppositions may be defined as follows:

(def. 3) A speaker S presupposes linguistically 'B' in uttering 'A' if:

(i) The representation of'B' follows from the application of implication rules on the representation of 'A'.

(ii) The speaker acts as if the representation of'B' were salient in the addressee's mental state.

In their turn, argumentative presuppositions do not come from the representation of a sentence via entailment. On the contrary, argumentative presuppositions are a whole vast and complex set of known facts about the world that serve as the basis for the construction ofarguments.

The distinction between linguistic and argumentative presuppositions cuts across the (alleged) distinction between Semantics and Pragmatics. As has been shown here, linguistic presuppositions come from the application on a conceptual structure of some rule of implication. However, the conceptual structure on to which sentences are mapped cannot be built up in the absence of contextual information. In other words, the inferential relations between natural language sentences have to be read off after a pragmatic process has built up the conceptual representation that maps onto the proffered utterance.

*F. García Murga**
The University of the Basque Country
Donostia - San Sebastián

NOTES

* I would like to thank Violeta Demonte for her comments on a previous draft and for bibliographical help. This paper has also benefited from its discussion in the ICCS'97. Obviously, mistakes are mine. This paper has been supported by a grant from the Government of the Basque Country.

[1] Focus has been defined as new information (Halliday 1967), as 'highlighting' of a linguistic expression (Bolinger 1985), as not c-construable (Cullicover & Rochemont 1983), and as a set of possible alternatives (Jackendoff 1972).

[2] This is the line I argued for in a previous analysis of sentence stress (García Murga, to appear).

[3] Even though very sketchily, such an analysis may be found in García Murga 1995a. Just to illustrate, what is claimed is that it is possible to foresee a theory that embraces the following data (capital letters stand for prominent stress):

(i) MURIEL voted for Hubert.
(ii) It was Muriel that voted for Hubert.
(iii) Even/Only Muriel voted for Hubert.

Sentences (i) to (iii) licences (iv):

(iv) Somebody voted for Hubert.

[4] See, however, section 2.1.

[5] Some problems in the analysis of 'even' (and, in general, of focal constructions) come from the selection of an appropriate variable. So, for instance, analyses done in line with Sperber and Wilson's Relevance theory have to include for each focused linguistic expression, a large set of specific rules. The account offered here (section 3) will not introduce any 'ad hoc' rule.

[6] See García Murga, 1995a, chapter 2. Basically, the most important problems that a logical definition of presuppositions cannot resolve are: (i) existential presuppositions in non-topic positions, (ii), existential and analytic sentences containing existential presupposition triggers, (iii) the cancellation of presupposition under negation, and (iv) the projection problem for presuppositions.

[7] It is probably more natural to think that sentence (4) is not a conventional implicature, but a generalised conversational implicature. Usually, given a scale, the derivation of the implicature is based on the first submaxim of Quantity:

> "Make your contribution as informative as is required (for the current purposes of exchange)" (Grice 1975 [Grice 1989: 26])

That is, using the metaphor of strength for defining the scale order, if the speaker uses a weak element in the scale, she implicates that the use of stronger elements will lead to a falsity.

8 See note 7 above.

9 "Make your conversational contribution such as is required, at the stage at which it occurs, by the accepted purpose or direction of the talk exchange in which you are engaged" (Grice 1975, [Grice 1989: 26])

10 Obviously, if the election is taken as carrying something positive for Hubert.

11 Other objections against Kay's entailment theory may be found in Kalokerinos 1995 and Anscombre & Ducrot 1983.

12 See Jackendoff 1983 and 1990, chapter 1.

13 There is a vast discussion between semantic approaches based in the combination of semantic features (taken as primitive elements) and semantic theories that take the lexical meaning itself as a primitive, unanalysable component such that meaning relations are defined via 'meaning postulates' (Carnap 1947). Even though the discussion is large and complex, a quite complete set of arguments for semantic decomposition may be found in McNamara & Miller 1989.

14 The analysis has been focused on the relations of implication between focal constructions and the (alleged) focal presuppositions. Thus, up to this point, little attention has been paid to the epistemological aspects of presuppositional phenomena. In previous works (García Murga, 1995 a, b), I have contended that linguistic presuppositions are taken as salient by the speaker. Saliency, then, is considered as a systematic feature of presuppositional constructions, i.e., of linguistic expressions that give rise to linguistic presuppositions. It is, then, reasonable to assume that saliency is somehow represented in conceptual structures. As a convention, I have used bold letters to indicate saliency. Then, a more accurate representation of (1) is:

(i)

$$\begin{bmatrix} \text{EVENT} \\ \text{CAUSE}\left(\begin{bmatrix}\textbf{[PLUR]} \\ || \\ \textbf{THING} \\ \text{MURIEL}\end{bmatrix}^{F}, \begin{bmatrix}\text{EVENT} \\ \text{HAVE}\left(\begin{bmatrix}\text{THING}\\ \text{HUBERT}\end{bmatrix}, \begin{bmatrix}\text{THING}\\ \text{VOTE}\end{bmatrix}\right)\end{bmatrix}\right)\end{bmatrix}$$

REFERENCES

Anscombre, J.C. and Ducrot, O., 1983: *L'argumentation dans la langue*. Bruxelles, Mardaga.

Bolinger, D., 1985: "Two views of accent", *Journal of Linguistics 21*: 79-123.

Carnap, R., 1947: *Meaning and Necessity. A Study in Semantics and Modal Logic*. Chicago, The University of Chicago Press.

Culicover, P.W. & Rochemont, M., 1983: 'Stress and Focus in English', *Language 59*: 123-165.

Ducrot, O., 1972: *Dire et en pas dire. Principes de Sémantique linguistique*. Paris, Hermann.

Fauconnier, G., 1975: "Pragmatic scales and logical structure", *Linguistic Inquiry 3*: 353-375.

Fraser, B., 1970: "An analysis of Even in English", in C. Fillmore and D.T. Langendoen (eds.), *Studies in Linguistic Semantics*. Holt, Rinehart & Winston, New York (pp. 141-180).

García Murga, F., 1995a: *Fundamentos conceptuales del fenómeno presuposicional*. Ph.D. dissertation. University of the Basque Country.

García Murga, F., 1995b: *On Linguistic Presuppositions*. Report No. ILCLI-95-LIC-4. Donostia: ILCLI.
García Murga, F. (manuscript): *A Procedural Analysis of Prosodically Driven Focal Presuppositions*.
Grice, H.P., 1975: "Logic and conversation", in P. Cole & J. Morgan (eds.), *Syntax and Semantics, vol. 9: Pragmatics* (pp. 113-127). Reprinted in H.P. Grice 1989.
Grice, H.P., 1989: *Studies in the Way of Words*. Cambridge, Massachusetts, Harvard University Press.
Halliday, M.A.K., 1967: "Notes on transitivity and theme in English", *Journal of Linguistics 3*: 177-274.
Horn, L., 1969: "A presuppositional analysis of only and even", Papers from the 5th Regional Meeting, Chicago Linguistic Society: 98-107.
Jackendoff, R., 1972: *Semantic Interpretation in Generative Grammar*. Cambridge, Massachusetts, MIT Press.
Jackendoff, R., 1983: *Semantics and Cognition*. Cambridge, Massachusetts, MIT Press.
Jackendoff, R., 1990: *Semantic Structures*. Cambridge, Massachusetts, MIT Press.
Jackendoff, R., 1996: "The Proper Treatment of Measuring Out, Telicity, and Perhaps Even Quantification in English", *Natural Language & Linguistic Theory* ***14***: 305-354.
Kalokerinos, A., 1995: "Even: How to make theories with a word", *Journal of Pragmatics* ***26***: 77-98.
Kay, P., 1990: "Even", *Linguistics and Philosophy* ***13***: 59-111.
Karttunen, L., 1974: "Presupposition and Linguistic Context", *Theoretical Linguistics 1*: 181-194. Reprinted in A. Rogers et al. 1977 (pp. 149-160).
Karttunen, L. & Peters, S., 1979: 'Conventional Implicature', in C.K. Oh & D. Dinneen (eds.), *Syntax and Semantics, vol. 11: Presupposition*. New York, Academic Press (pp. 1-56).
Kempson, R., 1975: *Presupposition and the delimitation of semantics*. Cambridge, Cambridge University Press.
König, E. (1991): *The Meaning of Focus Particles. A Comparative Perspective*. London & New York, Routledge.
McNamara & Miller, 1989: "Attributes of Theories of Meaning", *Psychological Bulletin* ***106***: 355-376.
Rogers, A. et al. (eds.), 1977: *Proceedings of the Texas Conference on Performatives, Presuppositions, and Implicatures*. Arlington, Virginia, Center for Applied Linguistics.
Sánchez de Zavala, V., 1991: Prolegomena to a (modest) pragmatical theory. Logic Seminar Report LPHS-EHU-02.2. University of the Basque Country.
Schmerling, S., 1976: *Aspects of English Sentence Stress*. University of Texas, Austin.
Sperber, D. & Wilson, D., 1986: *Relevance: Communication and Cognition*. Oxford, Blackwell.
Stalnaker, R., 1974: 'Pragmatic Presuppositions', in A. Rogers et el. 1977. (pp. 135-148).
Tancredi, Ch., 1992: *Deletion, Deaccenting and Presupposition*. MIT Working Papers in Linguistics.

TON SALES

LOGICS FOR MENTAL STATES

Abstract. Physical events are objective, meaningless, controllable by the laws of physics, and unconscious. Mental events are subjective, meaningful, intentional (i.e. they are "about" something), and conscious. Ordinary Science and Technology deal with physical events. Cognitive Science and Artificial Intelligence (AI) deal with intelligent behavior, which is usually linked to mental events. So, shall we give machines a mental structure? Conceiving our programs in terms of mental states makes our task clearer, but translating them into machine equivalents—and programming a computer from such a base—is no easy task, though some full-thrusted attempts have been made lately. We summarily describe them after introducing the mental state concept.

1. THE QUESTION OF MENTAL STATES

1.1. Physical Events vs. Mental Events: Intentionality

a) Physical objects are

— tangible

— objective

— meaningless[1]

— controllable by the laws of physics

— unconscious

[1] "Meaningless" = interpretation-independent

b) Mental events, on the other hand, are

— intangible

— subjective

— meaningful[2]

— intentional[3]

— conscious

[2] "Meaningful" = interpretation-sensitive

K. Korta et al. (eds.), Cognition, Agency and Rationality, 123-133.

[3] "Intentional" ≠ referring to a "purpose" or an "intention"

= that refers to something[4]

= that "it is about" something being referenced

[4] "Something" = any [possibly non-existent] object, without which the thing being referenced loses all possible meaning

Examples of **intentional** concepts (**mental states**):

Beliefs (and **Knowledge**)

Desires (or **Fears**), **Will**

Obligations

Intentions (= intended **Goals**)

(Here we exclude **Emotion,** though some authors like Bates consider them while others like McCarthy think they are superfluous—for an artificial entity—and even counterproductive.)

Intentionality is easily understood and used (⇒ **folk psychology**), but there are conceptual difficulties (see Horty and Shoham 1993, and Cox and Freed 1995) in *explaining* it as well as in *converting* it into useful Artificial Intelligence tools (⇒ the **grounding problem**).

1.2. The Philosophical Analysis of Intentionality

Philosophy has addressed the problem of intentionality by asking

Is there some physical correlate to intentional states?

In the *non-dualist* tradition favored by **computer scientists** the mainstream *materialist* tenet is aptly summarized in John Searle's phrase:

Brains cause minds

Thus (to most materialists): (*intentional*) **mental states** are reducible to (*physical*) **brain states.**

The nature of the connection is disputed. While some maintain that there can be no mental state without a proper brain, others (the "functionalists") hold that

A machine can also have mental states provided it is constructed to be functionally equivalent to a brain.

Anyway, many admit that the behavior of a machine can be explained by **intentional accounts,** i.e. attributing to it *beliefs*, *desires* and other mental states that humans have. Some theorists (notably Dennett, 1987) consider it natural while others (like McCarthy, 1979) accept it as a useful **metaphor** for brevity, or when there is no better **explanation** at hand.

In the first case (i.e. under the assumption that machines can *actually* behave **intentionally**), questions that arise naturally are: can such intentionality be adequately expressed in **logical** terms? can it be given an adequate semantics for "**grounding**" the mental states in some physical reality? (Such topics have been examined by McCarthy, 1995, Dennett, 1987, as well as Searle, Harnad and many others.)

1.3. Mental States and Consciousness

The relationship between **mental states** and **consciousness**, whose existence is obvious in the human context, poses some problems in the machine case: is consciousness *independent* of—or a *necessary* (or *sufficient*?) condition for—mental states? Is it accessible to scientific scrutiny?

The issue is relevant since **mental states** and **consciousness**—or their physical or functional correlates—have a real, **measurable effect** on the subject's actual **behavior**, and so allows us to predict it or anticipate it in some way.

2. ARTIFICIAL INTELLIGENCE AND THEORIES OF AGENTS

2.1. The 'Agents' Paradigm

A relatively recent trend in AI is centered on the **agent** concept (see e.g. the 1995 Russell & Norvig text).

An **agent** is a logically formalizable entity capable of

— **autonomy** (= non-continuous-dependence on the user)

— **initiative** (= use of own criterion or goals)

— **communication** (= interaction to share, signal, prepare for action, etc.)

— **cooperation** (allowing labor sharing or specialization)

— **reactivity** (= [reasonable-time-]response to detected change)

— **situatedness** (= adaptiveness to—and understanding of—the environment)

This approach is apparently more promising and flexible than the more traditional ones based on **knowledge-based systems**. It has been a chosen ground in which **intentional concepts** have been formalized and tried. (Wooldridge & Jennings, 1995 is a state-of-the-art report on AI agents.)

The so-called **agents' paradigm** links with rather recent concepts like DAI (Distributed AI) or MACC (Multi-Agent Cooperative Computing). It is somewhat richer than described above. For instance, it tends to attract people with problems like:

— modularity (= no central planning nor complex hierarchies)

— grounding (= contact with reality via continuous interaction)

— synergy (= total system more powerful than sum of parts)

(cf. **emergence** of complexity out of simple phenomena)

— parallelism (here built-in, and natural)

(Those qualities are often compared with the somewhat frustrating counterparts that ordinary AI systems tend to exhibit.)

2.2. How to Capture Intentionality

If we should treat **agents** and provide them **mental states** it may be better to distinguish a "weak" vision of what an agent is, by assuming it is (merely):

— autonomous

— sociable (= capable to interact through some linguistic means)

— reactive

— capable of having/showing initiative

To those basic traits we may require—and add, if we can or dare to—asking for those *stronger*, "*mentalistic*" characteristics from an agent (here we mention some of the more obvious or human-like):

— intentionality (in all or most of the above senses)

— reasoning ability

— goodwill (= no double standards or mixed intentions)

— veracity (= no possibility to mislead or deceive)

— rationality (= behave in the best way so as to achieve one's goals)

— character (= showing consistently own emotional states and style)

— commonsense (easy thing to recognize, hardest thing to define)

Intentional states had been already formalized in Logic by some authors, beginning with Hintikka, 1962 and Kripke, 1963 as the "**Logic of Belief and Knowledge**", a variety of modal logic which has also been used to model communication processes among agents (see e.g. Halpern & Moses, 1992). Other modal extensions, like **Temporal** and **Dynamic** Logics (see e.g. Goldblatt, 1992) are also relevant here because they are presently used to capture the dynamic aspects of intentional states.

2.3. Beliefs, Time, Communication and Intentions

Combining well-known logics like those modeling **belief, time** and **communication** is relatively straightforward, but **desires** and "**intentions**" (here meaning **goals**) are not so amenable to treatment.

Intention has been refashioned in a way so as to conform to criteria previously advanced by Bratman (1987). Syntheses have been attempted by different authors and have originated a variety of closely related or loosely equivalent logical formalisms called "**BDI logics**" (where "*BDI*" stands for "**Belief, Desire and Intentions**"). Here we just mention three formulations: those by

- Shoham (1993).
- Cohen & Levesque (1990).
- Rao & Georgeff (1991, 1993a, 1993b).

The description of these logics is preceded by the mention of a recent adaptation of the **Situation Calculus** of McCarthy to "*self-knowledge*" (in a robot) by McCarthy himself (1995).

3. LOGICS FOR MENTAL STATES

The four essential theories we describe here in summary form are **logical representation** formalisms. There are however other logical formulations such as Singh's (1991) or Werner (1990).

There are as well radically alternative explanations:

- R. Brooks' (or L. Steels')
  Reactive Systems (Brooks 1991, Steels 1988).
- Rosenschein & Kaelbling's
  Situated Automata (1986).

According to the first author, traditional AI has always been—deliberately and vocationally (cf. Newell's "physical symbol hypothesis")—**abstract symbol manipulation**, where the symbols were not actually connected to—or grounded on—reality. The systems were hierarchically controlled and increasingly complex (Brooks says they drown in "a sea of incrementalism") with no particular "intelligence" attachable to any particular module. Brooks rather proposes a new "situated" AI where elementary modules or agents, not particularly intelligent or complex, simply build up "intelligence" by coexistence and cooperation. Likewise, Steels has long emphasized the idea that representation in intelligent systems is more "analogical" than symbolic (discrete) or simply categorical (= previously categorized); he thinks that intelligence is rather the result of elementary operations of the diffusion/reaction type.

Here we choose to describe the more traditional approaches. "Traditional" in the sense that they are symbolic, logical and representation-based (albeit including a non-traditional **intentional** apparatus). The relevant information about the formalisms we describe below can be found in two technical reports by Horty and Shoham, and Cox and Freed (published in 1993 and 1995, respectively, by the American Association for Artificial Intelligence (AAAI) after two Spring workshop meetings), as well as in prior developments by the authors. Here we describe the main points.

3.1. McCarthy's Situation Calculus formulation

For McCarthy (1995) a robot is "a continuously acting computer program interacting with the outside world". What it "knows" about the world is encapsulated in logical sentences. A robot's "introspection" or "consciousness of self" is, for McCarthy, "a program having facts about its own mental processes and states of mind". So, the underlying assumptions are that (1) robots have, or can be considered to have, *mental states* and that (2) those mental states can be captured by the usual means of *logical representation* languages. Since consciousness of self is for McCarthy "essential for human level intelligence", it follows that a self-conscious robot is intelligent. For him intelligence does not require emotions

McCarthy (thinks that making robots sentient is not only unnecessary but even counterproductive). What *intelligence* requires is, for him: (1) *introspection* (= self-observation), (2) continuous *interaction* with the world, and (3) modeling *decisions to act* by strictly following the conclusions of a *logical reasoning* of some appropriate kind (for which the author proposes his own situation calculus, non-monotonic circumscription, context theories or other suitable logical formalisms).

To model self-knowledge McCarthy uses several propositional (meta)predicates —intentional propositional operators, actually—such as *holds*, *know*, *believe*, *learn* or *forget* (as well as *see*, which is an ordinary predicate taking an object, not a proposition, as the argument). The proposition may be of mental nature, and all predicates are unary except *holds*, which has a second parameter, the "situation" (as generally in McCarthy's *situation calculus*). He distinguishes propositions and *actions*. The latter are governed by particular predicates such as *occurs* and *result* (both binary, the second term being the situation) as well as *decide-to-do* or *intend-to-do* which are unary. Though *occurs* yields a proposition *result* gives a state-transition function, *decide-to-do* is an action and *intend-to-do* a "fluent" (a kind of event that persists unless something changes it). With those elements, McCarthy can express propositions such as

holds (know p, s), holds (know not know p, s) or holds (know Telephone Mike, s)

(where the capitals in *Telephone* and *Mike* mean that we are not referring to the denoted objects but to their concepts, which are *not* exchangeable modulo identity). Other expressions he is able to use are

occurs (a, s) or *result (a, s)*, which are self-explanatory.

He can also express and fix some rules governing the intended system (the robot agent), such as:

holds (know p, result (learn p, s))

occurs (p, s) $\rightarrow$ *holds (F not know, s)*

holds (F p, s) $\rightarrow$ *holds (p, next (p, s))*

occurs (forget p, s) $\rightarrow$ *holds (p, next (p, s))*

occurs (decide-to-do a, s) $\rightarrow$ *holds (intend-to-do a, s)* , or

occurs (see o, s) $\rightarrow$ *holds (F knows did see o, s)* ,

which are transparent enough to be read unaided. The *not* is ordinary negation, while *F*, *next* and *did* are modal operators (the first means "sometime in the future").

With those ingredients McCarthy proposes to formalize, manipulate and explore the knowledge a robot can have of itself. He envisions this logical extension as the starting point of a theory of the effect of mental states, a "Logic of mental events" with which we may in the future be able to prove that a given sequence of actions yields (or not) the purported goal the robot set itself to attain.

McCarthy tentatively adds a further predicate *hope* with the idea that if *result* equals *hope* things are apparently going well. If on the contrary hopes generally do not coincide with results, that should signal a strategy revision. The author also proposes to evaluate the robots's ignorance (its non-knowledge) by allowing it to know that it does *not* know something: the first method that comes to mind, that of finite failure, is dismissed by McCarthy, who proposes instead a second-order procedure based on Gödel-Cohen's proof ofthe relative consistency of set theory and the continnum hypothesis in a method which is strongly reminiscent of the author's second-order circumscription formula.

3.2. BDI Logics: the Shoham formulation

The *BDI Logics* ("BDI" for "Belief", "Desire" and "Intention") are the simplest implementation of *mental attitudes*. The first two operators *B* and *D* refer to the agent's informative and evaluative states, while *I* refers to decisions made by the agent corresponding to his optimal behavior when adapting to resource limitations).

Though originally put forward and written already in 1990, Yoav Shoham proposes in 1993 formalizing intentional concepts like *beliefs*, *obligations*, *decisions* (or *choices*), *commitments* and *capabilities* through epistemic modal operators complemented with a normal Kripke semantics, a temporal logic and some regulating axioms and interdefinitions. He defines an *agent* as an autonomous high-level symbolic entity composed of mental states that are characterizable by modal concepts such as *belief* or *knowledge*, provided they are precisely defined in logical terms, communicable (to other agents) and complemented by other intentional and action operators like *obligation*, *choice*, *decision* or *commitment* inside the frame of Shoham's ambitious "agent-oriented programming" project. After discarding *choice* as a primitive metapredicate in favor of *decision*—which he further reduces to *obligation*, a *decision* being defined as obligation to oneself—he chooses to treat *time* as a stamp or tag appended to an action (paralleling McCarthy's *situation* parameter), he then treats events as *actions* and defines a short list of primitives: *belief*, *obligation* and *commitment*, for which he provides a rigorous semantics. He can thus express situations like those ones:

holding (robot, cup)^t (where the *t* superindex is the time-stamp)

$B^t{}_a\varphi$ (meaning that agent *a* believes the proposition φ at *t*). Example:

$B^3{}_a\, B^{10}{}_b$ *like (a,b)* expresses that *a* believes that seven instants later *b* will believe "*like (a,b)*".

$OBL^t{}_{a,b}\varphi$ (meaning that *a* has an obligation towards *b* over the fact φ)

Here Shoham adds that $DEC^t{}_a\varphi$ (the decision to do φ) is, by definition, the same as $OBL^t{}_{a,a}\varphi$ (one's obligation towards oneself over φ) and that $DEC^t{}_a\varphi$ implies—but not vice versa—$B^t{}_a\varphi$ (i.e. if one decides to do something he believes it will take place).

As to *capabilities*, Shoham distinguishes between "can" ($CAN^t{}_a\varphi$) and "being able" ($ABLE_a\varphi$). The latter is rather like an immediate version of the former, as the interdefinition shows: $ABLE_a\varphi = CAN^{t(\varphi)}{}_a\varphi$; thus, *CAN* is the latent capability (as believed by a) to perform φ at some later time $t(\varphi)$ which is the time in which φ occurs. Thus,

$ABLE_{robot}\ open\ (door)^t = CAN^t_{robot}\ open\ (door)^t$.

Shoham defines some expectable and exigible properties in his BDI logic. Here we enumerate some:

— *internal consistency*: for any agent a and time t the sets

$\{\varphi \mid B^t_a\varphi\}$ and $\{\varphi \mid OBL^t_{a,b}\varphi$ for some $b\}$ are consistent.

— "*good faith*": for any t, a, b and φ;

$OBL^t_{a,b}\varphi \rightarrow B^t_a(\varphi \wedge ABLE_a\varphi)$

— *introspection*: for any t, a, b and φ;

$(OBL^t_{a,b}\varphi \leftrightarrow B^t_a OBL^t_{a,b}\varphi)$, and $(\neg OBL^t_{a,b}\varphi \leftrightarrow B^t_a \neg OBL^t_{a,b}\varphi)$,

— *persistency* of mental states (assumed by default):

$A[t] \rightarrow A\ [t']$ whenever $t < t'$, where $A\ [t]$ is a statement about a mental state (at time t)

Shoham's piece has at least two added advantages: he lists the authors and efforts who came before him attempting more or less the same (and he does not forget to mention who was probably the first one of them, R.C. Moore and his pionerering 1980 work on on "reasoning on knowledge and action") and compares them with his own work. He also conceives his as a three-stage project: a *logic* of mental states, an interpreted *programming language* and a sort of compilation he calls *agentification* (that he leaves undeveloped).

3.3. BDI Logics: the Cohen & Levesque formulation

In parallel formulation to that of Shoham's, and—like him—attempting to apply classical Speech-act Theory to intentional actions by closely following Bratman's criteria (Bratman 1987), Philip Cohen and Hector Levesque (1990) built in 1990 and afterwards a modal logic with standard Kripke semantics plus (linear) time based on the modal *belief*, *goal* and *intend* operators plus the temporal *happens* and *done*.

For Bratman (1987), *rationality* is composed of *belief*, *desire* and *intention*, but the latter is, though related, not reducible to the former two, because *intending* is not simply acting intentionally but coordinating the agent's *plans*, which may be the result of competing desires.

Unless Shoham, Cohen & Levesque distinguish between *events* and *actions*. The non-temporal modal operators they define and provide with a firm Kripkean grounding are those three: *BEL* (*belief*), *GOAL*, and *HAPPENS*. They add a temporal or action-related operator *DONE* plus the usual temporal model operators □ (meaning as usual "always") and ◇ ("eventually [in the future]"), for which they provide a (linear) time semantics, to which they add secondary operators like *LATER* and *BEFORE*. They also distinguish ordinary goals from "persistent goals" (they notate by *P-GOAL*) and between what they call a "fanatical" commitment to do action *a* (that they notate by $INTEND_1$) as distinct from a more relativized one $INTEND_2$. The definitions ofthe first one is as follows:

$$INTEND_1\ x\ a = (P\text{-}GOAL\ x\ [DONE\ x\ (BEL\ x\ (HAPPENS\ a))\ ?;\ a])$$

where it is apparent the dynamic-Logic style notation the authors use.

Shoham and Cohen-Levesque are well aware of some (minor) divergences between their systems. Apart from the use of time-stamps against modal temporal operators, they also note that Shoham's *decision* does not involve any motivation or desire by the agent, it merely signals the actions to which the agent feels commited, while Cohen and Levesque's *goal* or *choice* does include the agent's desires. Moreover, for Cohen and Levesque belief implies choice:

$BEL_a p \rightarrow GOAL_a p$, in contrast with Shoham's formula shown above indicating the contrary (but, as the authors remark, they are rather technical consequences of the type ofKripke semantics each chose at design time).

3.4. BDI Logics: the Rao & Georgeff formulation

A. Rao and M. Georgeff (1991, 1993a, 19993b) build a formalism they call "agent-oriented situated systems" which is broadly equivalent to the two previously mentioned BDI logics. They do so by modeling the intentional *belief*, *desire*, *intention* and *commitment* concepts plus a branching time semantics and an emphasis on practical aspects such as agent specification, design and verification.

As Cohen and Levesque in 1990 they also try to conform to Bratman's criteria (1987), which include the so-called *asymmetry thesis*: "It is irrational for an agent to intend to do *a* and at the same time to believe the he will not do *a*, but it is rational for an agent to intend to do *a* and not to believe the he will do *a*". That amounts to saying that *irrationality* is having beliefs inconsistent with intentions, but irrrationality does not imply having incomplete beliefs over intentions (on the contrary, this can be perfectly rational).

Cohen and Levesque (1990) deal with Bratman's thesis but in Rao and Georgeff's opinion they do no solve satisfactorily its side-effects. They refashion Bratman's *asymmetry thesis* as follows (see Rao and Georgeff 1991):

1) No $\models INTEND\varphi \wedge BEL\varphi$

2) There is a model in which $M \models INTEND\varphi \wedge \neg BEL\varphi$

To this formulation they add this *side-effect-free principle*: There exists a model *M* in which

$$M \models INTEND(\varphi) \wedge \Box BEL(\Box(\varphi \rightarrow \psi)) \wedge \neg INTEND\psi$$

as well as this *non-transference principle*: There exists a model *M* in which

$$M \models BEL\varphi \wedge \neg GOAL\varphi$$

With these three formal principles they characterize the necessary conditions for an agent's *rationality*. They are then free to add what they call weak axioms ofrealism:

$\models GOAL\varphi \rightarrow \neg BEL\neg\varphi$

$\models INTEND\varphi \rightarrow \neg GOAL\neg\varphi$

$\models INTEND\varphi \rightarrow \neg BEL\neg\varphi$

but not this one (that we mentioned earlier) that Cohen and Levesque obtain by postulating a linear semantics:

$\models BEL\varphi \rightarrow GOAL\varphi$

Upon these pemises Rao and Georgeff later built up a whole agent-oriented architecture of the "BDI Logic" type (see Rao and Georgeff 1993a) along parallel lines with Cohen and Levesque but with a branching time semantics.

Rao and Georgeff's modal language consists of the three classical BDI operators *BEL*, *DESIRE* and *INTEND*, plus the following branching time operators: X (next), *UNTIL*, F (eventually, i.e. sometime in the future), E (at some path in the future), E_ξ (some committed path ξ in the future), G (any time in the future, i.e. always), A (in all paths in the future, i.e. inevitably), A_ξ (in all committed paths ξ in the future), as well as F^∞ (infinitely often) and G^∞ (nearly always).

In the semantics that Rao and Georgeff define, a Kripke *frame* is a septuple $<W,T,R,B,G,L>$ where T is the time, R is the time accessibility relation, and B, G, and I are the accessibility relations between beliefs, desires and intentions, respectively. Each world w in W is the triple $<T_w, R_w, L_w>$, where T_w is a subset of T, R_w is the restriction of R to the points T_w of T, and L_w is the function assigning time points in T_w to propositions. In this semantical context belief, for one, is thus defined:

$M,w,t \models BEL\varphi$ iff $M,w',t \models \varphi$ for any φ such that $w, t, w' \in B$

When Rao and Georgeff speak of *situated systems* they mean that their behavior —as ruled by its mental attitudes—are *reactive* and *goal-directed*. Such architecture, it is suggested, is better implemented by assuming the *agent A* is *situated* in a specific environment E and definable as a map $A: S \times E \rightarrow B$ where S are the mental states and B are the rational behaviors b, so that an agent is defined by the function $b(s,e)$.

Ton Sales
Universitat Politècnica de Catalunya
Spain.

REFERENCES

Bratman, M. E.: *Intentions, Plans, and Practical Reason*. Harvard University Press, 1987.

Brooks, R. A.: Intelligence without representation, *Proc. IJCAI*, Sydney (1991)

Cohen, P. R. & Levesque, H. J.: Intention is choice with commitment, *Artificial Intelligence* ***42*** (1990), 213-261

Cox, M. & Freed, M., eds.: *Representing Mental States and Mechanisms*. (Papers from the 1993 AAAI Symposium). AAAI Technical Report SS-95-05 (1995)

Dennett, D.: *The Intentional Stance*. MIT Press (1987)

Goldblatt, R.: *Logics of Time and Computation*. CSLI (1992)

Halpern, J. Y., & Moses, Y.: A guide to completeness and complexity for modal logics of knowledge and belief. *Artificial Intelligence* ***54*** (1992), 319-379

Hintikka, J.: *Knowledge and Belief*: Cornell University Press (1962)

Horty, J., & Shoham, Y., eds.: *Reasoning about Mental States: Formal Theories and Applications*. (Papers from the 1993 Spring Symposium). AAAI Technical Report SS-93-05 (1993)

Kripke, S.: Semantical analysis of modal logic. *Zeitschr. f. Math. Log. u. Grundl. der Math.* ***9*** (1963), 67-96

McCarthy, J.: Ascribing mental qualities to machines, in: Ringle, M., ed. *Philosophical Perspectives in Artificial Intelligence*. Harvester Press (1979)

McCarthy, J.: Making robots conscious of their mental states. In (Cox, M. & Freed, M., eds. 1995) (1995), 89-96

Rao A. S. & Georgeff, M. P.: Asymmetry thesis and side-effect problems in linear time and branching time intention logics. *Proc. IJCAI*. Sydney (1991), 498-504

Rao A. S. & Georgeff, M. P.: A model-theoretic approach to the verification of situated reasoning systems. *Proc. IJCAI*. Chambery (1993a), 318-324

Rao A. S. & Georgeff, M. P.: Verification of agent-oriented situated systems: a model-theoretic approach. In (Horty, J., & Shoham, Y., eds. 1993) (1993b), 115-124

Rosenschein, S. J., & Kaelbling, L. P.: The synthesis of digital machines with provable epistemic properties. *Proc. TARK*. Morgan-Kaufmann (1986)

Russell, S., & Norvig, P.: *Artificial Intelligence: A Modern Approach*. Prentice Hall (1995)

Shoham, Y.: Agent-oriented programming, *Artificial Intelligence* ***60*** (1993), 51-92

Singh, M. P.: A logic of situated know-how. *Proc. AAAI* (1991), 343-348

Steels, L.: Steps toward common sense. *Proc. ECAI*. Munich (1988)

Werner, E.: What can agents do together: a semantics of co-operative ability. *Proc. ECAI*. Stockholm (1990)

Wooldridge, M, & Jennings, R. N.: Intelligent Agents: Theory and Practice (report submitted Oct. 1994 to the *Knowledge Engineering Review*, revised Jan. 1995)

SUSANNE KALENKA, NICHOLAS R. JENNINGS

SOCIALLY RESPONSIBLE DECISION MAKING BY AUTONOMOUS AGENTS

Abstract. Most autonomous agents are situated in a social context and need to interact with other agents (both human and artificial) to complete their problem solving objectives. Such agents are usually capable of performing a wide range of actions and engaging in a variety of social interactions. Faced with this variety of options, an agent must decide what to do. There are many potential decision making functions which could be employed to make the choice. Each such function will have a different effect on the success of the individual agent and of the overall system in which it is situated. Therefore, this paper examines agents' decision making functions to ascertain their likely properties and attributes. A framework for characterising social decision making is presented and a socially responsible decision making principle is proposed which enables both the agent and the overall system to perform well. This principle is illustrated, and empirically evaluated, in a multi-agent system for unloading lorries at a warehouse.

1. INTRODUCTION

Intelligent agents are a new paradigm for developing software applications. More than this, agent-based computing has been hailed as "a new revolution in software" (Ovum 1994) and it has been predicted that "agents will be pervasive in every market by the year 2000" (Janca 1995). Consequently, agents are the focus of intense interest on the part of many sub-fields of computer science and artificial intelligence. Agents are being used in an increasingly wide variety of applications, ranging from comparatively small systems such as email filterers, to large, complex mission critical applications such as air traffic control. Despite this apparent diversity, in all cases the key abstraction used is that of an *agent*. Although there is much debate about exactly what an agent is (see Franklin and Graesser 1996 for a discussion), we consider the following to be necessary conditions for agenthood: (i) *autonomy*—can act without the direct intervention of others and has control over its own actions and internal state; (ii) *responsiveness*—can react in a timely fashion to environmental changes; (iii) *proactiveness*—can take the initiative where appropriate; and (iv) *social ability*—can interact to complete its problem solving and to assist others (Wooldridge and Jennings 1995).

The increased autonomy afforded by the agent paradigm means that an agent's decision making function is central to the success of any application. This function takes a subset of the agent's beliefs—including, for example, its current state, the

K. Korta et al. (eds.), Cognition, Agency and Rationality, 135-149.

state of the environment, and the state of other agents—and determines the course of action the agent should follow. Almost by definition, this function operates with a partial view of the world (Lesser and Corkill 1987) and, because of the inherent interdependencies between the agents (Davis and Smith 1983), the choices it makes affect not only itself but other agents in the environment. Given this situation, it is important to consider the following fundamental questions: what is a good decision making function? and what measure should be used to rate goodness? Present work in multi-agent systems can be divided into two broad camps (figure 1). The reductionist view (e.g. Erman and Lesser 1975, Lesser and Corkill 1983) is concerned with building effective overall systems using the notion of interacting agents, while the constructionist view (e.g. Ferber and Drogul 1992, Fischer *et al.* 1996, Overgaard *et al.* 1996, Steels 1989, Wavish and Graham 1996) considers interacting agents as a given and is concerned with determining what sort of overall system emerges from them.

Figure 1. Extant approaches to social decision making

	Reductionist View	**Constructionist View**
Motivation	Devise a system of interacting agents that work together to solve a common problem.	Devise agents which interact to further their own needs.
Measure of Goodness	Overall system performance.	Performance of individual agents.
Agents' Decision Making Function	Benevolence—accept all requests made.	Individual utility maximisation.
System Coherence	Carefully engineered by single design team.	Emerges out of interplay between agents.
Main Drawback	Fails to fully exploit concept of autonomous agents—too much system level design.	System behaviour defined through human-intensive refinement of individual agents.

Whilst both of these approaches have enabled useful applications to be developed, it has been predicted that agents will have the greatest impact in complex industrial and commercial applications such as process control, telecommunications management, business process management, air traffic control, manufacturing, and information management (Ovum 1994). In such systems, what is required is the ability to exploit the conceptual power of autonomous agents (as in the constructionist view), but to ensure the overall system performs in a coherent manner (as in the reductionist view). Given these contradictory demands, we believe the best means of building socially coherent multi-agent systems is to endow the individual autonomous agents with greater social awareness. This awareness enables the agents to explicitly consider the effects of their actions on the wider community. Given this information, a decision function can then be designed which enables agents to exploit interactions with others for their own gain, but which means that they are sometimes willing to do things for the greater good (to improve system coherence). Here the term *social responsibility* is used to denote such agents.

The aim of this paper is, therefore, to elucidate the decision making principles of socially responsible agents. It is hypothesised that such agents are the best means of designing multi-agent systems in which a balance needs to be struck between the needs of the individual agents and the needs of the overall system (henceforth the society). To this end, section 2 presents an informal framework for characterising social decision making. Section 3 uses this framework to identify a principle of socially responsible decision making and identifies three specific socially responsible decision making functions (socially self-interested, helpful and cooperative). Section 4 illustrates the use of these functions in a multi-agent system for unloading lorries at a warehouse and presents an empirical evaluation of their effectiveness. Section 5 discusses related work and, finally, section 6 outlines some issues which require further investigation.

2. A FRAMEWORK FOR CHARACTERISING SOCIAL DECISION MAKING

Most work on decision making functions for autonomous agents concentrates on making individually rational choices (e.g. Doyle 1992, Russell and Wefald 1991, Wellman 1993). This work either uses a decision theoretic notion of utility maximisation (given the probability of being in a state after executing an action and a rating of the desirability of that state, choose the action which maximises the expected utility (Doyle 1992) or a more intuitive description such as Newell's Principle of Rationality (if an agent has knowledge that one of its actions will lead to one of its goals, then it will select that action (Newell 1982). In either case, however, the decision function is solipsistic—there is no consideration of the impact of actions on other members of the society and no notion of doing anything other than maximising the agent's own gain. It would be possible to manipulate the agent's utilities (or goals) so that they incorporate a measure of social awareness, but this would simply be hiding the underlying principles behind the numbers.

In order to be explicit about the underlying principles of social decision making, we note that both the agent who performs an action and the society in which that agent is situated can be affected by the execution. This affect can be beneficial (a positive utility value), detrimental (a negative utility value) or indifferent (a utility value of zero). Also individual and society benefit are orthogonal measures. In more detail:

> **Individual Benefit (a, S, α)**: the benefit[1] agent a, situated in society S, obtains for performing action α is a combination of the benefit attributed solely to the action executor (*agent sole benefit*) and the executor's share of the benefit the society obtains when one of its members executes a (*agent share benefit*). For example, if a team of agents is searching for a particular document on the world wide web, then the system may be organised such that the individual who actually finds the document receives 50% of the customer's payment (agent sole benefit) and the remaining 50% is split evenly between the team (in which case the finder also receives some portion of this 50% as its agent share benefit).

> **Society Benefit (a, S, α)**: the benefit society S obtains when one of its members (agent a) executes action α. This is composed of the executor's agent share benefit (as above) and the benefit accrued by those members of S who do not actually execute the action (*other share benefit*). For example, if there are four agents searching for a document in the above scenario and the benefit is distributed evenly amongst the group, then the agent which finds the document receives 25% ofthe team fund and the other three agents receive 75%.

In seeking to design socially responsible agents, neither individual, nor society benefit alone is adequate (the former may preclude doing actions for the good of the society and the latter may preclude the agent from furthering its own needs). Therefore a hybrid measure of benefit, termed *joint benefit* (Jennings and Campos 1997), is adopted. Joint benefit incorporates both the individual and the societal perspectives:

> **Joint Benefit (a, S, α)**: a combination of the individual benefit agent a obtains for executing action α and the benefit obtained by society S in which agent a is situated. Clearly the agent share benefit should not be counted twice, so joint benefit is a combination of the agent's sole benefit, the agent's share benefit, and the others' share benefit.

With this framework in place, it is now possible to return to the decision making function. All decision making functions perform two primary roles. Firstly, they identify the set of candidate actions that the agent may consider performing. For example, a self-oriented agent will only consider actions which have a positive individual benefit, an altruistic agent will only consider actions which have a positive value for other share benefit, and so on. Secondly, the function defines a preference ordering over the candidate actions. For example, a selfish self-oriented agent may rank actions in decreasing order of individual benefit, whereas a more socially aware self-oriented agent may rank them in decreasing order of individual benefit and where there is a choice in terms ofdecreasing society benefit.

Figure 2: Spectrum of Decision Making Functions

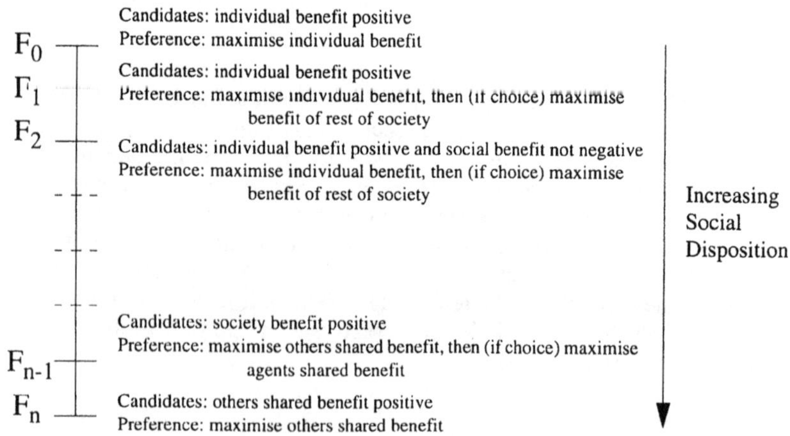

From this structure, a whole spectrum of potential decision making functions can be observed (figure 2)—ranging from the purely selfish (F_0) to the purely altruistic (F_n). Our concern, therefore, is with identifying the subset of this function space which leads to socially responsible behaviour. This activity is performed in section 3.

3. A SOCIALLY RESPONSIBLE DECISION MAKING PRINCIPLE

Rather than seeking to posit a single socially responsible decision making function, we feel it is more useful to identify the underlying principle(s) on which such functions are founded. This principle can then be instantiated in a variety of ways depending on the relative importance of the individual and the society in a given application. Thus our approach is in line with Newell's view of rationality (identifies a broad candidate set and no preference ordering), rather than the decision theoretic view ofrationality (fixed, single-point candidate and preference structure).
In seeking to strike a balance between the needs ofthe individual agents and those of the society, we use joint benefit (section 2) as our foundational basis. Thus, the guiding principle for socially responsible agents can be defined in the following way (Jennings and Campos 1997):

> **Principle of Social Rationality:** If a member of a responsible society can perform an action whose joint benefit is positive, then it may select that action.

In this work, we concentrate on three socially responsible decision making functions—namely, socially self-interested, helpful and cooperative (Kalenka and Jennings 1995). These functions represent a reasonable spread of the more individually-centred socially responsible alternatives and therefore maintain many of the individualistic aspects oftraditional autonomous agents.

- *Socially self-interested:* represents the most individualistic type of responsible agent. Considers candidate actions which have a positive joint benefit (as per the Principle of Social Rationality) and which have a positive individual benefit. These agents primarily concentrate on their own actions, but ensure they are not overly detrimental to the society. They differ from purely self-interested functions in that they do not simply execute actions without any regard for their affects on others.

- *Helpful:* permits non-detrimental actions which are of no direct benefit to the agent, but which are beneficial for the society as a whole. These agents consider candidate actions which have a positive joint benefit (as above), but also consider actions which have no positive individual benefit. The latter stipulation enables agents to choose actions which benefit the society but which are not individually beneficial.

Both ofthe aforementioned decision making functions consider only isolated actions (i.e. without regard to other potentially related actions). However in many cases it is important to consider combinations of actions executed by groups of agents

(Ketchpel 1993, Shehory and Kraus 1995). Here we deal only with the simplest case in which there are two agents each executing a single (related) action. Consideration of such cooperative actions means a further extension to the agents' decision making functions—now they need to include both the action they are to execute and the action the other agent is to execute.

- *Cooperative*: considers candidate actions which have a positive joint benefit and pairs of actions (one executed by the local agent and one by the other team member) which, when taken together, have a positive individual benefit. Thus, for example, one of the actions may have an individual benefit which is negative, but this loss is compensated for by the execution of the other agent's action. This means an agent can execute a personally detrimental action for the good of the society (joint benefit positive) as long as it receives sufficient individual benefit from the accompanying team action.

It is clear that the cooperative decision function requires some form of social commitment (Castelfranchi 1995, Jennings 1993) to be made between the two agents. Without such a commitment, agents will not consider individually detrimental actions since they cannot be guaranteed to recoup their loss by the execution of the subsequent action. The nature of the social commitment can vary in scope (from being valid for just the current pair of actions, up to an ongoing partnership to help one another out whenever necessary (Kalenka and Jennings 1995)) and the convention under which it can be terminated (from one agent opting out because it has a better offer, to one requiring mutual acceptance of the termination (Jennings 1993)). In this work, an ongoing commitment structure is adopted because it provides a safe, long-term basis for entering into cooperative problem solving. It works by the agents committing to support one another for periodically repeating time intervals. These intervals are called *support-duty* when one agent has to support the other and *support-right* when one agent can demand support from the other. Each agent alternates between support-right and support-duty intervals. During an agent's support-duty interval it supports the other agent if it is requested to do so. When an agent's support-right interval arises it can demand up to an equivalent amount of support from the other agent as gave in its last support-duty interval. Thus, the lower the amount of support requested, the lower the support that must be given. The social commitment terminates when one of the agents does not request any support in its support-right interval. This conceptualisation means there is no predetermined end point for a commitment and in some cases it may continue indefinitely. At each support-right interval a cooperative agent has to determine how much it should force the other agent to support it as this directly affects the time it may be unable to work on its own goals at the next support-duty interval.

4. A MULTI-AGENT SYSTEM FOR UNLOADING LORRIES IN A WAREHOUSE

This section describes how the socially self-interested, helpful and cooperative decision making functions can be used in a practical application context—in this case for designing a multi-agent system for unloading lorries at a warehouse (section 4.1). The effectiveness of the decision making functions in this domain are then evaluated in a series of experiments (section 4.2).

4.1 Applying the socially responsible decision making functions

Lorries arrive randomly at a warehouse laden with goods which require unloading. The warehouse has a fixed number of unloading bays which each hold one lorry at a time. Upon arrival, lorries go to the nearest free bay. Lorries have an associated time by which they would like to be discharged (their desired time). Additionally, the warehouse tries to ensure that all lorries are processed by some maximum time (T_{max}). Since the aim of this work is to illustrate social problem solving behaviour, rather than to develop a real world solution, we make a number of simplifying assumptions: all lorries arrive with the same load; the time it takes one fork lift truck (agent) to unload a lorry is T_{max}; unloading time is directly proportional to the number of agents servicing a lorry (two agents will do it twice as fast as one); and there are as many fork lift truck agents unloading the lorries as there are bays in the warehouse.

In our system, each agent is responsible for dealing with a particular unloading bay. Each agent receives some sole benefit for ensuring the lorry at its assigned bay is processed by T_{max}. Moreover, if the lorry is processed in a desired time which is less than T_{max} the agent responsible for that bay receives further sole benefit. The society as a whole receives benefit proportional to the percentage of lorries that are processed by T_{max} and the percentage processed by their desired time (where this is less than T_{max}). This benefit is split between the agents which assist others with their unloading tasks and is in proportion to the amount of assistance provided.

In more detail, let L_i be the lorry to be discharged, A_i be the agent responsible for L_i, and T_i be L_i's discharge deadline. Four distinct cases need to be considered:

- $T_i > T_{max}$: A_i can discharge L_i on its own and it has some spare time (T_{max} - T_i) in which it could support other agents. (Lorry is *not-time-dependent*).
- $T_i = T_{max}$: A_i can discharge L_i on its own. (Lorry is *time-dependent-alone*).
- $T_i < T_{max}$: A_i needs support from other agents if it is to meet L_i's desired deadline. Without any support discharge will take T_{max}. (Lorry requires *time-dependent-support*).
- T_i unknown: No L_i has arrived at bay i, hence A_i is free to support other agents at least until T_i is defined.

If all agents used the socially self-interested decision function, then the multi-agent system could be guaranteed to meet its objectives as long as no lorry needed to be discharged in a time less than T_{max} (i.e. there are no time-dependent-support lorries). However, this situation can be improved by the agents adopting the helpful decision making function. In this case, agents could support one another when they have a not-time-dependent lorry or when they have no lorry at their bay. Helpfulness would ensure the agents are more heavily utilised and that more deadlines less than T_{max} are met (simply because there is more agent problem solving power available in the system). The situation can be enhanced still further by the addition of the cooperative decision making function. For example, consider the case where two lorries arrive simultaneously. Lorry L_1 has the desired time of T_{max} (it is time-dependent-alone) and lorry L_2 the desired time of T_{max} / 2 (it requires time-dependent-support). If A_1 and A_2 cooperate, as defined in section 3, A_1 could assist A_2 for the first T_{max} / 2 units of time (meaning L_2 is discharged in time) so long as A_2 agreed to assist it for the second T_{max} / 2 units of time (meaning L_1 is also discharged in time). With only responsible or helpful agents, both lorries would be discharged in time T_{max} meaning L_2's desired departure time is not satisfied. In the cooperative case, the two agents make a social commitment to support one another while either of them are in danger of not fulfilling their responsible tasks. Thus if a lorry arrives at A_2's bay while it is assisting A_1 with L_1, then A_1 must commit itself to help A_2 clear the new lorry, and so on. In the worst case, new lorries, with deadlines less than or equal to T_{max} plus the time the responsible agent is unavailable to start work, may arrive continuously at A_1's and A_2's bays while they are supporting one another. Thus their ongoing commitment will continue until: (i) no lorries arrive before the agents fulfil their duties at the other's bay; (ii) one of the lorries has a sufficiently long discharge time; or (iii) either agent receives the necessary support from some other helpful agent.

4.2 Experimental results

These experiments provide an empirical assessment of the performance of the three different types of socially responsible decision making function in the warehouse unloading application. In the particular scenario considered here, there are ten agents and ten unloading bays and a 50% chance that a new lorry will arrive at an empty bay on a particular simulation cycle. New lorries arrive in the ratio of 1 not-time-dependent lorry to 1 time-dependent-alone lorry to 3 time-dependent-support lorries. In a given experiment, all the agents have the same decision making function —thus they are all socially self-interested, or all helpful, or all cooperative.

The basic socially responsible hypotheses being evaluated in these experiments can be stated in the following manner:

- as the agent's decision making function is made more socially aware (socially self-interested to helpful to cooperative), the performance of the society will improve.
- as the agent's decision making function is made more socially aware, the performance of the individual agents will not deteriorate significantly.

The first experiments consider the success rate of the different types of agents over the three types of lorry (subsequent experiments examine effectiveness by lorry type). Figure 3 (and all subsequent graphs) shows the success rate at each of the 10 bays separately and the average over all the bays. The average value can be considered as the performance of the overall society. Each bar shows the percentage of lorries at a particular bay which have been discharged in less, equal and greater than the desired discharged time. Figure 3 demonstrates that the helpful decision function generates a significant improvement over the socially self interested function —the percentage of lorries which miss their desired time drops from 58.6% to 34.7%. This improvement is achieved by making better use of spare capacity in the system and ensuring that lorries are not needlessly processed before their desired time (for which no credit is given). The cooperative decision making function gives a still greater improvement (3.1%) over the helpful decision function although this was not as much as had been expected. The reason for the smaller than expected improvement is because the social commitment structure used in this scenario binds pairs of cooperating agents together for prolonged periods of time. This means there are comparatively few new opportunities to offer assistance.

Figure 3: Success Rate over all lorry types.

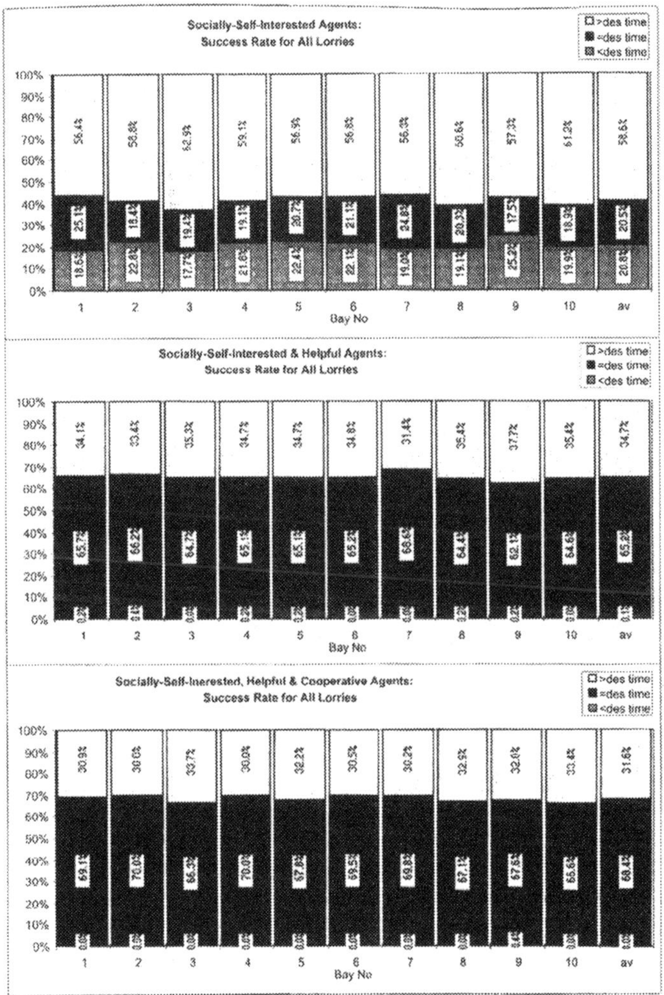

To provide a more detailed breakdown ofthese figures, an analysis of the success of the different functions for the different types of lorry was undertaken. For both the socially self-interested and the helpful cases, all of the lorries that were either not-time-dependent or which were time-dependent-alone were processed by their desired time (graphs not shown). The big improvement, as might be expected, occurs with respect to those lorries who require time-dependent-support. With the socially self-interested decision function no lorry meets its desired time, but with the helpful function 40.7% ofthe lorries meet their desired time (figure 4).

Figure 4: Success of Socially Self-Interested and Helpful Decision Functions for Time-Dependent-Support-Lorries

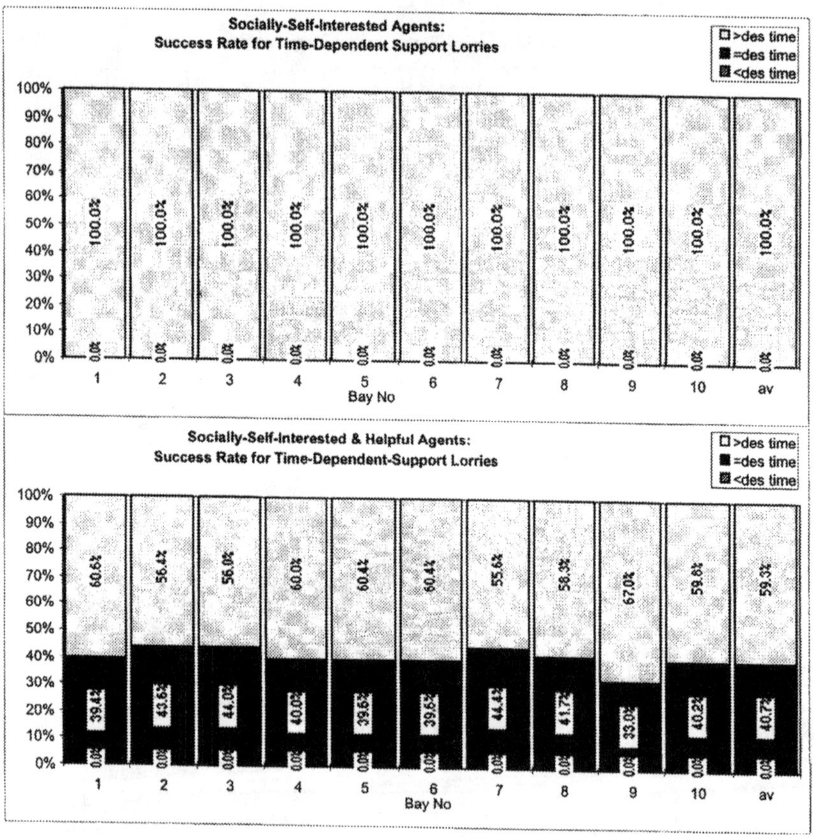

The cooperative decision making function reveals a further 5.5% improvement for time-dependent-support lorries over its helpful counterpart (figure 5, second graph). Thus indicating that the greatest benefit ofmore socially aware functions is likely to occur in cases in which there is a greater social dependence between the participating agents. However, it is interesting to note that not all of the time-dependent-alone lorries are processed by their desired time (figure 5, first graph) (cf. the socially self-interested and helpful decision functions). All of these lorries could have been processed in time had the agent responsible for their bay taken a more self-oriented view. It is precisely because such agents enter into social commitments with one another that in some cases their individual performance suffers—they are doing actions for the greater good of the society even if they turn out to be individually detrimental some of the time.

Figure 5: Success of the Cooperative Decision Function for Time-Dependent-Alone and Time-Dependent-Support Lorries

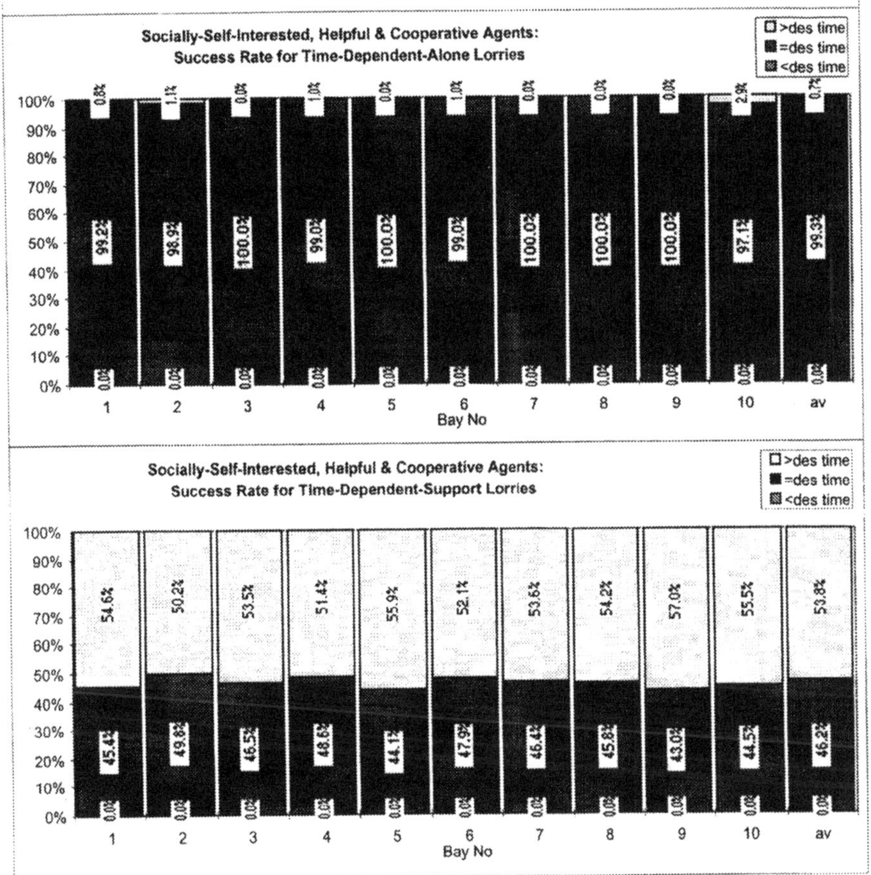

5. RELATED WORK

A range of disciplines are concerned with the theoretical and experimental study of social interactions between autonomous agents. However a comprehensive analysis of this work is beyond the scope of this paper. Therefore, we concentrate on that research which is closest to our work as it is described in this paper.

Castelfranchi (1990) considers the motivation for social action in multi-agent systems from the perspective of social science. He identifies the notions of individual and social power as important influences on the nature and type of interactions which occur. One agent has the power to influence the social problem solving behaviour of another when the latter is dependent on it. In this context, dependence means that an agent cannot complete one of its goals without assistance. With this view, agents assist one another (undertake helpful or cooperative actions) because of their interdependence, not because they wish to achieve greater levels of system coherence. This differing emphasis means that agents need to reason about dependency networks (Sichman *et.al.* 1994) rather than about societal utilities and also that agents are still essentially individual utility maximisers.

Cesta *et al.* (1996) explore the social problem solving behaviour of groups of agents with various interaction attitudes related to their degree of self-sufficiency and the degree to which they are willing to give help. In particular, they study the performance of social agents (which give and take support) in the presence of agents which try to exploit them. Their results show that, depending on the threshold at which agents are willing to be helpful, social agents can tolerate exploiter agents without a severe decrease in system performance. This is encouraging for our socially responsible societies as it indicates that they cannot be readily exploited by outside agents which may have different decision making principles.

Sen (1996) considers interactions between different types of self-interested agents. In particular, he considers the effects, on both individual and system performance, of including agents which receive help from others and which do, and do not, reciprocate. He concludes that reciprocal behaviour can improve the performance of the individuals and of the overall system in which they are situated. Moreover, agents which reciprocate helpful behaviour can both approach optimal global behaviour and resist exploitation by selfish agents. Again these results indicate the advantages of taking a socially responsible stance.

Finally, Marsh (1992) discusses trust as a computational concept which autonomous agents can use when deciding with whom to interact (in our context this equates to deciding who to help or who to cooperate with). Factoring such a concept into our social decision making would allow agents to use functions with an increased social disposition (towards F_n in figure 2) with agents for which they have a high degree of trust and more self-oriented functions (towards F_0 in figure 2) with those agents for which they have a low trust value.

6. CONCLUSIONS AND FUTURE WORK

This paper presented a framework for characterising social decision making in multi-agent systems. It provided a typology of the types of benefit associated with action execution and identified social commitment as a key supporting structure. The Principle of Social Rationality was put forward as a foundational basis for designing autonomous agents which strike a balance between their own problem solving objectives and the needs of the society. Three socially responsible decision functions which adhere to this principle were described and their success evaluated in a multi-agent system for unloading lorries in a warehouse. These experiments highlight the value, in terms of overall system performance, of making agents' decision making functions more socially aware. However they also indicate that, in some cases, individual performance may suffer when trying to enhance system coherence.

There are a number of aspects of this work which require further investigation. Firstly, it is assumed that all the agents are trustworthy and able to make accurate predictions about their level of commitment to one another. But what will happen to individual and system performance if some of the agents renege upon their commitments? (either deceitfully to try and exploit others or because they were simply unable to make accurate predictions about their resource availability). What sorts of mechanisms can agents put in place to protect themselves against such free-riding agents? Secondly, the present social context is represented at a coarse level of granularity—involving just the individual agent and the whole society. However in many cases, there are several social groupings within the society with various strengths of relationships between their members. Given this situation, how can a more differentiated framework for social decision making be defined? Finally, determining utilities for social actions consumes the agent's resources. In most cases these resources are limited and so it may not always be possible to determine all their values. Thus, the agent may need to take a resource-bounded view on social rationality (Hogg and Jennings 1997)—meaning the agent's decision making function is only an approximation to the ideal social rationality outlined here. In such cases, agents need to be designed which are able to manage their computations such that the more resources they have at their disposal the closer they perform to the ideal.

*Susanne Kalenka**
Nicholas R. Jennings
Queen Mary and Westfield College. University of London.
United Kingdom

NOTES

* This work has been supported by an EPSRC studentship and by a Drapers Company scholarship.

[1] In this context, the term "benefit" covers positive, negative and indifferent utility values.

7. REFERENCES

Castelfranchi, C., 1990: Social Power: A Point Missed in Multi-Agent DAI and HCI, in Y. Demazeau and J. P. Müller (eds.), *Decentralized AI* , Elsevier, pp. 49-62.

Castelfranchi, C., 1995: Commitments: From Individual Intentions to Groups and Organisations, Proc 1st Int. Conf. on Multi-Agent Systems, San Francisco, USA, 41-48.

Cesta, A., Miceli, M., and Rizzo, P., 1996: Help Under Risky Conditions: Robustness of the Social Attitude and System Performance, Proc 2nd Int Conf on Multi-Agent Systems, Kyoto, Japan, 18-25.

Davis, R., and Smith, R. G. 1983: Negotiation as a Metaphor for Distributed Problem Solving, *Artificial Intelligence* ***20***, 63-109.

Doyle, J., 1992: Rationality and its Roles in Reasoning, *Computational Intelligence* ***8*** (2), 376-409.

Erman, L. D., and Lesser, V. R., 1975: A multi-level organisation for problem solving using many diverse cooperating sources of knowledge, Proc. Int. Joint Conf. on AI, Stanford, CA., 483-490.

Ferber, J., and Drogul, A., 1992: Using Reactive Multi-Agent Systems in Simulation and Problem Solving in Distributed Artificial Intelligence: Theory and Praxis, in N. M. Avouris and L. Gasser (eds.), Kluwer Academic Publishers, pp. 53-80.

Fischer, K., Müller, J. P., and Pischel, M., 1996: Cooperative transportation scheduling: an application domain for DAI, *Int. Journal of Applied Artificial Intelligence* ***10*** (1), 1-33.

Franklin, S., and Graesser, A., 1996: Is it an Agent, or just a Program, Proceedings Third International Workshop on Agent Theories, Architectures and Languages, Budapest, Hungary, 193-206.

Hogg, L. M., and Jennings, N. R., 1997: Social Rational Agents— Preliminary Thoughts, Proc. of Second Workshop on Practical Reasoning and Rationality, Manchester, UK.

Janca, P. C., 1995: Pragmatic Application of Information Agents, BIS Strategic Report.

Jennings, N. R., 1993: Commitments and Conventions: The Foundation of Coordination in Multi-Agent Systems, *The Knowledge Engineering Review* ***8*** (3), 223-250.

Jennings, N. R., and Campos, J. R., 1997: Towards a Social Level Characterisation of Socially Responsible Agents, *IEE Proceedings on Software Engineering* ***144*** (1) 11-25.

Kalenka, S., and Jennings, N. R., 1995: On Social Attitudes: A Preliminary Report, Proc. First Int. Workshop on Decentralised Intelligent Multi-Agent Systems, Krakov, Poland, 233-240.

Ketchpel, S., 1993: Coalition Formation Amongst Autonomous Agents, Proc 5th European Workshop on Modelling Autonomous Agents in a Multi-Agent World, Neuchatel, Switzerland, August 25-27.

Lesser, V.R., and Corkill, D.D., 1987: Distributed Problem Solving, in S. C. Shapiro (ed.), *Encyclopedia of Artificial Intelligence*, John Wiley and Sons, pp. 245-251.

Lesser, V.R., and Corkill, D.D., 1983: The Distributed Vehicle Monitoring Testbed: A Tool for Investigating Distributed Problem Solving Networks, *AI Magazine*, Fall, 15-33.

Marsh, S., 1992: Trust and Reliance in Multi-Agent Systems, Proc 4th European Workshop on Modelling Autonomous Agents in a Multi-Agent World, Rome, Italy.

Newell, A., 1982: The Knowledge Level, *Artificial Intelligence* ***18***, 87-127.

Overgaard, L., Petersen, H. G., and Perram, J. W., 1996: Reactive motion planning: a multi-agent approach, *Int. Journal of Applied Artificial Intelligence* ***10*** (1), 35-51.

Ovum Report (1994): Intelligent agents: the new revolution in software.

Russell, S., and Wefald, E., 1991: *Do the right thing*, MIT Press, Cambridge Mass.

Sen, S., 1996: Reciprocity: a Foundational Principle for Promoting Cooperative Behaviour among Self Interested Agents, Proc 2nd Int. Conf. on Multi-Agent Systems, Kyoto, Japan, 322-329.

Shehory, A., and Kraus, S., 1995: Task Allocation via Coalition formation among autonomous agents, Proc. 14th International Joint Conference on Artificial Intelligence, Montreal, Canada.

Sichman, J. S., Conte, R., Demazeau, Y., and Castelfranchi, C., 1994: A Social Reasoning Mechanism Based on Dependence Networks, Proc 11th European Conf on AI, Amsterdam, The Netherlands, 188-192.

Steels, L., 1989: Cooperation between distributed agents through self organisation, *Journal of Robotics and Autonomous Systems*.

Wavish, P., and Graham, M., 1996: A situated action approach to implementing characters in computer games, *Int. Journal of Applied Artificial Intelligence* ***10*** (1), 53-73.

Wellman, M. P., 1993: A Market-Oriented Programming Environment and its Application to Distributed Multi-commodity Flow Problems, *Journal of Artificial Intelligence Research* 1, 1-23.

Wooldridge, M. J., and Jennings, N. R., 1995: Intelligent Agents: Theory and Practice, *The Knowledge Engineering Review* ***10*** (2), 115-152.

ROSARIA CONTE

DIVERSITY IN RATIONALITY. A MULTI-AGENT PERSPECTIVE

1. LIMITS OF RATIONALITY

The marriage between AI, on one hand, and economic and strategic rationality, on the other, is unavoidable: there are too strong "elective affinities". For many purposes and domains, the AI adoption of the economic paradigm (Wellman 1995) or of the game-theoretic apparatus (Gmytrasiewicz 1995), is both motivated and productive. However, these paradigms are often imported a-critically, as mere techniques or instruments. The debate occurring in the original disciplines about the foundations of these approaches has been ignored. One should import from other disciplines not only the solutions but also the related background problems. This should be true also within AI, if this is considered as a science rather than a mere technology (engineering). Although the significance of the economic and strategic approach for (D)AI is undeniable, the theory-and-design of intelligent autonomous agents, as well as the theory-and-design of MA systems, require models of rationality and models of social action to go far beyond these paradigms.

Here, the merits of the economic theory of rationality will only be re-stated. Indeed, in the first section of the paper, the merits of Game Theory (GT) will be shown to be particularly relevant for DAI and MAS. Afterwards, a number of criticisms to economic decision theory that have an impact on its significance for modelling intelligent social/individual action, will be examined. These criticisms differ from both the most classical ones (no computational limitations, completely specified preferences, complete knowledge) and others more recent developed in decision theory (Castelfranchi et al. 1992). One of them, namely, economic reductionism, will be analysed at some length. The interpretation of rationality allowed by economics will be challenged. Economic rationality will be argued to differ from rationality tout court and be unnecessary for artificial agents. Also the concept of rationality provided by decision theory (SEU) is neither necessary nor sufficient for modelling agents and MAS. It says nothing about the agents' goals and preferences; and it is not sufficient for modelling both individual and collective decision processes. In general, the notion of rationality needs a deep reconsideration, what is beyond the scope of this paper. The amendments that will be proposed essentially amount to introducing diversity in the rationality paradigm. Several types of diversity will be suggested, from substantive diversity —a eterarchic view

K. Korta et al. (eds.), Cognition, Agency and Rationality, 151-174.

of motivations as opposed to a monarchic view (profit), in which agents are supposed to be indifferent to the various qualitatively different motives of the agents— to architectural diversity: in humans, for example, at least a further strategy (goal-based rationality) exists, in which utility is subordinated to goal-satisfaction, rather than vice versa.

2. DRAWBACKS

Aside the problems already addressed in the relevant literature (no computational limitations, multiple equilibria, etc.), other aspects of rationality theory need reconsideration.

2.1. A monarchic view of rationality

Rationality is seen as a "monarchic" notion (profit) indifferent to the various qualitatively motives of the agents.

A monarchic architecture

What rational decision theory has to say about the agent architecture and its mind (content) is very poor. Its instrumentalist view gives us some sort of "monarchic" mind, driven by one and unique goal: *utility maximisation,* though bounded by cognitive limitations. Real agent motivations are irrelevant: what matters is not "what" to do and "why", but just "how much" one can gain in one or other activity. What counts is not the agent's commitment to a given intention (and need satisfaction) but its investing as much as possible in the most profitable activity. There is in fact only one *need* to be satisfied: utility.

Multiple equilibria: when monarchy leads to anarchy

As a mere *model of the individual's choice* instrumental rationality is not so good:

> "the principal (problem) has arisen because there seem to be rather a lot of settings where there are *multiple* rational expectations equilibria". With game-theory, "the problem is exactly the same as the one which has surfaced with respect to multiple rational expectations equilibria (...). An individual's choice is underdetermined when there are several Nash or perfect equilibria because *any of the Nash/perfect equilibria will satisfy the conditions of instrumental rationality* and knowledge provided it is selected by others." (Hargreaves-Heap 1993, p.75-77; our italics).

Therefore, the economic approach and GT lead to some troubles also as decision criteria implemented in the agent's architecture; often, they cannot give any indication about how to choose among different alternatives. This is quite clearly expressed by Gmytrasiewicz (1995):

> "There are .. important issues for which the field of game theory does not seem to provide a satisfactory answer. One of them has to do with the case of multiple equilibrium points existing in a game. The question is, which

one should be chosen...? For example, ... should one prefer equilibrium points that have a better payoffs, or the ones that are safer?" (p.I/42).

Collective decision making

Of course, what is true for individual decision making is also true for *collective decision making*, with some serious additional problems:

(a) "A related question is: how do the players achieve beliefs that are *common knowledge*, and .... converge on a particular equilibrium in the first place? This issue is problematic, particularly on view of the results due to Halpern and Moses (1984)" (Gmytrasiewicz 1995, p.I/42);

(b) What may be considered as rational from the strictest individual viewpoint may prove irrelevant or dangerous from the global point of view. This is so in many respects, ofwhich only a few have been accounted for so far.[1] *Utility theory and GT do not help us so much manage conflicts between the global or collective utility of a population of agents or of a team, and the local, personal utility of the members*: should we subordinate the individual (or sub-agent) to the collectivity (and why should this sacrifice be accepted by the individual?) Or should we subordinate collective advantage to individualistic motives? Again, in designing MAS, one will have to take such a decision.

2.2. Game Theory troubles with Multi-agent systems: the notion of "cooperation"

GT and social action

The basic notion of "cooperation" in GT, i.e. the "cooperative" behaviour in the Prisoner Dilemma (PD), is undefined and implicit: it is just an *intuitive* notion — based on a prototypical *scenario of interpretation* of the payoff matrix— plus the real formal properties of the matrix, which neither justify the term "cooperate" nor support such notion.

GT agents are unusual cognitive agents: their mental representations are mutilated. They have beliefs (or better knowledge), but do not have goals, i.e. mental representations of a worldstate wanted by the agent. Instead, they have utilities, preferences. Goals *control* and *guide* action thanks to repeated tests of the action's results against the current worldstate, determine the action search and selection, determine its success (and thereby its reward) or failure, etc. Our claim is that *without an explicit goal-representation, a theory of action and in particular of "social action" is impossible.*

What is in fact a social action? We define a social action as an action that achieves a social goal. In other words, at some level in the chain of a social action's goals one ofthese two alternatives is mentioned:

a) an *action* performed by another agent y, or

b) some *mental attitude* of y (goal, belief, or emotion).

y must be mentioned as an autonomous agent (endowed with mental attitudes and the capacity to pursue its own goals; for a discussion of this notion, see Conte & Castelfranchi 1995 Ch. 3) in one of x's goals.

Cooperation vs bargaining: goal-interdependence vs social costs

The formal construction of the PD game is rather complex and subtle but allows neither the mechanism of cooperation nor that of social action at large to be fully understood. The PD-game structure endeavours to incorporate a social network, namely a network of interdependence among two or more agents, in a frozen pay-off matrix. Two orders of problems deserve attention here:

(a) the nature, character, and source of *interdependence*, and consequently;

(b) the implicit characterisation of sociality in terms of individual *costs*.

Our claim is twofold. First, given its notion of interdependence, the PD game structure does not allow the *variety* of positive social actions (exchange, help, cooperation) to be accounted for, and is bound, again, to a *monarchic* view of sociality; secondly, and more specifically, given its implicit characterisation of positive social action in terms of specific social costs, the PD-game structure does not allow cooperation as such to be accounted for. In other terms, PD-game structure could not grasp the variety of pro-social phenomena. And the only one which is admissible within the framework of the PD-game structure, could not have been cooperation, but at most a forerunner of exchange.
Undoubtedly, a PD-game structure is *an interdependence structure*, where the players' outcomes depend on each other's moves. But what does this *exactly* mean? What type of interdependence is constructed, and what is its source? The outcome interdependence is but an interdependence of utility values: each player's utility maximisation depends on the other's move. A fundamental character of the utilitarian nature of interdependence is that it is exclusively *formal*:

(a) it is *instrumentalist* rather than substantialist, that is to say, it is destitute of any explicit content;

(b) it is *built-in*, hardwired into the structure of the game rather than emerging from some lower-level phenomena (for example, the agents' qualitative properties, capacities, goals, resources, etc.);

(c) it is exclusively *quantitative* rather than qualitative.

In order to fully understand *utilitarian* interdependence, one should compare it with another, *goal-based* notion of interdependence (cf. Conte & Castelfranchi, in press;

Conte & Castelfranchi 1995; Sichman et al. 1994). According to the latter, agents depend on one another when they need others' help to achieve their goals. Unlike the former, it is based on goals and means-goal relations; since goals are explicit representations, goal-based interdependence is *substantialist* (goals are determinate); *emerging* from lower-level phenomena (the social structure of interdependence is derived from, and built upon, the agents' properties, their goals and capacities); primarily *qualitative* (since goals are symbolic representations), which are assigned a quantitative value (on the basis of means-end relations).

The immediate consequence of GT characterisation of interdependence is that it leads again to a *monarchic* view of social action: what game-theorists call Cooperation is the only form of sociality allowed by the structure of a PD-game. This structure does not attain the variety of pro-social interaction, ranging from unilateral self-interested help (which occurs when x lends her room-mate her car tonight in order to get rid of him and have the whole flat for herself) to exchange (bilateral help based on expectation of being reciprocated), and from this to cooperation (common activity based on agents' complementarity).

The PD-game structure and the monarchy of bargaining

In *explicit* terms, social action is defined by game-theorists as any move which takes into account possible moves of the opponent players (coordination). *Implicitly*, and more meaningfully, a positive social action ("Cooperation") is an action which leads the agent to sustaining an intrinsically *social cost*. In other words, there is no cooperation without a penalty for the cooperative agent. This implicit conceptualisation of social action ties it up to the paradigm of bargaining. It characterises sociality as a costly and dangerous move, in which agents punish and reward each other at the same time, in which there can be no benefit without costs, in which agents face each other each trying to get away scot free. It is a view of sociality as a necessary evil, where agents are fundamentally opponents (as game-theorists indeed define them). Such a view is deeply engrained in the utilitarian philosophy. Actually, it is the only possible view of sociality if one takes a fundamentally utilitarian, that is to say formal in the sense previously defined, notion of interdependence. But it is by no means the only possible view of sociality. *Why should a "cooperative" move be by definition less convenient than other moves? Why should it necessarily have an additional "social" cost?*

There is another view of sociality, based on a substantialist, emerging, and qualitative notion of goal-interdependence, therefore based on the agents' goals and goal-relations, which does not define pro-social action as an action with inherent social costs. Therefore, such a view is not bound to the monarchic law of negotiation. As was shown elsewhere (Conte & Castelfranchi 1995), a goal-based notion of interdependence leads to a variety of forms of social actions, including but not reduced to bargaining. Agents' *reciprocal dependence* (that is, their depending on each other to achieve different goals) is shown to lead to bargaining, that is, to each agent accepting a cost (to pursue the other agent's goal), in order to obtain some benefit from the other's achievement. But goal-based dependence allows another structural relation among the agents to emerge, namely *mutual dependence*, that is to say, the agents' depending on each other to achieve *one and the same goal*. This category of analysis is not allowed to emerge from a utilitarian interdependence, because by definition it requires an explicit, substantialist notion of utility. In a word, it requires goals. Mutual dependence leads to cooperation, that is to say, to the common achievement of a common goal, to the execution of a multi-agent plan. *Now, such a notion of cooperation is not an intrinsically costly social action.* It is a co-interested, rather than self-interested form of cooperation: since both agents aim at achieving the same end, and to that end they are complementary, they need that each accomplishes its share of the common plan. Such a notion allows rational choice to be reconciled with a cooperative, pro-social action even in one-shot interactions, and this is precisely because it allows for a view of sociality as an immediate advantage, rather than a necessary evil.

3. AMENDMENTS: DIVERSITY IN RATIONALITY

In this paper, it is argued that:

(a) rationality is a relative and contextual notion: there is not an absolute rationality;

(b) there is more than one type of rational agent;

(c) there is more than one type of rational decision-making.

Quite on the contrary, a notion of rationality is called for which does not rule out diversity.

There are many senses in which diversity must and can be integrated with a theory of rationality:

(a) by introducing *substantial differentiation*: qualitative heterogeneity among individual agents' goals; to do this, a goal-based rather than preference-based, view of endogenous motivations should be provided. The difference between goals and preferences is fundamental in agent theories and architectures (for a for a discussion about the difference between goals and preferences and the advantages of the former notion over the latter, see also Gmytrasiewicz 1995). The essential difference between them is in the qualitative vs quantitative characterization: while preferences are quantitatively defined, goals are symbolic, qualitative notions. Unlike the former, they allow for agents to be heterogeneous.

(b) By introducing *architectural differentiation*: agents do not only differ in the contents of their goals but also in their architectural properties. Therefore, they do not only have different goals, but also different decision-making rules. (Notice that in evolutionary game-theory, different strategies for interaction are allowed but the principle of utility maximization in decision-making is never questioned!) In our view, a further principle (*goal-based rationality*) exists, in which *utility is subordinated to goal-satisfaction*. rather than viceversa.

(c) By allowing for a *variety of social action* to be described and predicted: in terms of rational decision theory applied to social settings only the alternative between defection or cooperation is possible. But social life is interspersed with different types of pro-social action, from *influencing*, to *exchange*, to *cooperation*. It is argued that utility cannot actually account for such a variety, while a qualitative notion can.

(d) By admitting for a *context-bounded* notion of rationality, such that different contexts call for different rational strategies;

In the remaining of the paper, our preliminary contribution to these objectives — especially to the first three— will be illustrated. Some aspects of a model of intelligent (social) action will be described, and applied to partially different computational tools developed within the framework of Multi-Agent Systems and Social Simulation.

4. A MULTI-AGENT APPROACH

In our view, a (social) *action* is defined in terms of (social) goals (cf. Conte & Castelfranchi 1995).

A *goal* is a symbolic representation of the world, which the agent wants to be true. A goal is not yet an action. Goals are ordered according to precedence or preference relations; but a goal is not primarily defined by its order position. A social action is defined as aimed to modify another agent's mind.

A (social) *agent*, in a quite elementary sense, is defined as characterised by goals, actions, (planning) knowledge and resources. The notion of agent here referred to is defined as: (a) an ideal-type construct—not to be confused with the subjective, idiosyncratic individual of differential psychology; and (b) an AI-based notion, where the attention is drawn on the whole process that leads a system to acting, and on its internal makeup, i.e., the internal regulatory mechanisms and representations allowing a system to act adaptively in its environment.

4.1. Substantial diversity and interaction variety

In Sichman et al. 1994, a computational instrument (DEPNET), calculating the network of dependence relations among agents in a common environment, has been developed, based on a very simple agent architecture, consisting of agents' goals, actions, and resources. In Conte & Sichman 1995, and Conte et al. 1996 DEPNET has been applied to allow a complex structure of agents' interdependencies to emerge, and as a consequence, a variety of rational social (inter)actions to be predicted (influencing, exchange, and cooperation).

Below, we will describe a theory of dependence as presented in Sichman et al. 1994 on the basis of a pre-existing model developed by Castelfranchi et al. 1992, and Conte & Castelfranchi 1995.

4.1.1. The Dependence model

Social agents are plunged into a network of social relationships. The focus is on the agents' mental states, namely their goals. Social networks are here seen as patterns of relationships holding among the goals and actions of a given set of agents.

The most fundamental relationship among agents' goals and actions is *social dependence* (Castelfranchi et al. 1992), where one agent needs the action of another to achieve one of her goals.

The three basic notions of the social dependence theory are *social market*, *dependence relation* and *dependence situation*. We will present here only the first two; the last one is available in Sichman et al. 1994; Conte & Sichman 1995; Conte 1996.

The social market

We will call a social market, or a market for short, any aggregate of agents where the value of a single agent's resources depend on the wants and needs of the others. In other words, in a social market, agents reach their goals thanks to what they have to "sell". The general principle for achieving one's goals is that you-have-what-I-need-*and*-I-have-what-you-need.

The social market consists of a data structure composed by:

(a) the set of *goals* each agent wants to achieve,

(b) the set of *actions* she is able to perform,

(c) the set of *resources* she controls and

(d) the set of *plans* she has. A plan consists of a sequence of actions with its associated resources needed to accomplish them.

However, an agent may have a plan whose actions or resources do not necessarily belong to her own set of actions or resources, and therefore she may *depend on others* in order to carry on a certain plan, and achieve a certain goal.

An *entry* corresponding to an agent ag_j has respectively:

- the set of goals,
- actions,
- resources and
- plans the external observer believes ag_j has.

By, *resources*, we mean, concrete objects that may be required by performing actions. For the time being, we will conceive of resources as both non-consumable and re-usable (for example a pair of scissors is a resource for cutting a piece of cloth). In future developments of the model, both constraints will actually be dropped.

Finally, actions are costly. In this paper, however, and again for the sake of simplicity, action costs are assumed to be equal.

Dependence relations

We define the notions of autonomy and dependence as follows.[2]

An agent ag_i is *a-autonomous* (action autonomous) for a given goal g_k, according to a set of plans P_{qk} if there is a plan that achieves this goal in this set and every action appearing in this plan belongs to her own set of actions $A(ag_i)$ In other terms, an agent is a-autonomous if she is endowed with all the actions involved in at least one of the plans that achieves her goal: if her set of plans is non-empty, but none of those plans is exhausted by her actions, the agent is not a-autonomous.

Analogously, we define the notion of *r-autonomy* (resource autonomy).

Finally, an agent ag_i is *s-autonomous* (social autonomous) if she is both a-autonomous and r-autonomous for this goal.

On the other hand, if an agent does not have all the actions (or resources) to achieve a given goal, according to a set of plans, she may depend on the others for this goal.

An agent ag_i *a-depends* (action-depends) on another agent ag_j for a given goal g_k, according to a set of plans P_{qk} if ag_i has g_k in her set of goals, she is not a-autonomous for g_k and there is a plan in P_{qk} that achieves g_k where at least one action used in this plan is in ag_j's set of actions $A(ag_j)$

In a similar way, we have defined the notion of *r-dependence* (resource-dependence).

Finally, an agent ag_i *s-depends* (social-depends) on another agent ag_j if she either a-depends or r-depends on this latter.

In Conte & Castelfranchi 1995, some notions describing multi-party relations have been formalised.

AND-dependence occurs when one and the same agent s-depend on a bunch of others for achieving one[3] of her goals.

Conversely, the agent a_i is in a relationship of *OR*-dependence with regard to a set of agents Ag_j when she s-depends on any of them for a given goal. This equals to saying that each member of Ag_j has at least one action sufficient for ag_i's goal to be achieved. Any member of the set Ag_j is sufficient but unnecessary for ag_i's goal. For example, in order to have information about how to draft a given application to some international research program I need either one or other of a bunch of colleagues more experienced on the subject matter than myself. OR-dependence provides the dependent agent with a number of alternative ways to achieve her goal, among which she shall choose the most convenient. The number of alternatives amounts to the number of agents contained in the set Ag_j.

Finally, if a set of agents Ag_j s-depend on ag_i for one and the same goal, a relation of *CO*-dependence occurs between Ag_j and ag_i. The latter is a contended for social resource. In other words, ag_i happens to find himself at the centre of a dependence network.

From *CO*-dependence, a number of phenomena may follow. In particular, ag_i may be compared with others according to the number of agents CO-depending on him. We will call *social utility* a value expressing the extent to which each agents is useful for the others, measured in terms of their CO-dependence. CO-dependence forms the grounds upon which the relative social utility of the agents in a dependence network can be evaluated and compared.

4.2. Architectural diversity

There is *more than one type of rational agent* (a claim which is now shared by several authors; cf. Hargreaves-Heap 1993) argument that some games are solved by players "departing" from a presumption of common rationality); and moreover, that there is *more than one type of rational decision-making.* Actually, there are several strategies of decision-making depending on the criteria built into the deciding system. Let us examine at least one general criterion, compared with the classical utility maximisation, namely: the *goal-based strategy.*

4.2.1. Goal- vs utility -based strategies

While in the utility-based principle *goal-satisfaction is subordinated to maximising utility;* and the real goal is just to maximise utility,

$Max\ U(X_1, ..., X_n; I_t)$
where Max stands for the maximisation principle; $X_1, ..., X_n$, *the set of alternatives, with their relative probabilities; and* I_t *the available information;*

in the *motivation-based rationality* the system tries to achieve the most-valued goal; *utility is subordinated to goal-satisfaction*: motivations are not equivalent / interchangeable. This strategy consists of choosing in favour of the most-valued goal,

$Most\text{-}V(G_1,...,G_n; K_t; A_1,...,A_n)$
where Most-V stands for a choice function; $G_1,...,G_n$ *for a set of alternative goals;* K_t *the knowledge available to the agent at any given time (including knowledge about the world, as well as planning knowledge);* $A_1,...,A_n$ *the set of actions available to the deciding agent.*

Therefore, given a set of alternative *active* goals, the agent will decide to achieve the most-valued one, provided it is believed to be achievable, convenient, and unfulfilled. Suppose, x has a set of ordered (active) goals: $g_1>g_2>...>g_n$. Suppose that x prefers to go out with Jim over dining with John, but finds the latter alternative more convenient as for the ratio between cost and incomes (gain). The two principles allow for two different predictions: the utility principle leads to choosing John, while the goal principle enables x to choose Jim.

Several *basic* theoretical questions arise here: what is meant by a goal, what is the difference between goals and utilities, or more precisely, between goals and preferences, which are the foundations of utility theory? What is the difference, if any, between maximising one's satisfaction and maximising one's utility? etc. In the following section, the latter question will be addressed to some further detail

4.2.2. *The interplay between goal-satisfaction and utility-maximization*

The distinction introduced above is rather more complex than might appear. Indeed, the two strategies in question are actually intertwined in the real matters. In the following, a variety of substrategies will be analytically derived from the interaction between these main strategies.

(a) *Blind hedonism (or, addiction)*. The most-valued goals will always be achieved: given $g_1>g_2>...>g_n$, g_1 will always be chosen even if it has negative consequences on other goals

(b) *Prudent (or, self-defensive) hedonism.* The most-valued goals will be achieved only if they are not self-destructive.

(c) *Hard (or obstinate) hedonism.* If the most-valued goal cannot be achieved, exit.

(d) *Next-best hedonism.* If the most-valued goals cannot be achieved, then choose the next-best.

(e) *Failed hedonism* (or, utilitarianism by indifference). Utility is maximised only when hedonism is inapplicable. If there are no (significant) discrepancies in the intensity or urgency of desires, goals, needs, etc., then resort to utility.

(f) *Mixed: hedonistic choice + utilitarian execution.* The most urgent goal is chosen, but it is applied in a utilitarian way. The most-valued goals will be achieved in the most economic way. (The other way around does not make much sense)

(g) *Opportunistic hedonism.* As many goals as possible will be pursued, provided the most-valued one can be achieved.

Let us just stress a couple of very important differences between the two heuristics.

(a) The principle of maximising utility renders the agent indifferent to both the content and the context of her activity! Since what counts is the final utility obtained, it should be totally indifferent how that value is realised. The application of a general principle of utility maximisation which is indifferent to the content and the context of decision is essentially counter-intuitive, allows for too complex and useless decision-making, and finally may easily lead to deadlocks.[4] On the contrary, a mechanism of decision-making which is based upon goal satisfaction rather than utility maximisation seems to avoid some of these consequences: goals are not fungible. Indeed, the decision-making space is constrained by at least two mechanisms:

(i) situated activation: not all possible profitable investments (activities/ goals) are considered, but the choice is only among those agent's goals that are active in the specific situation the agent is involved in. We believe that this situated rationality is quite different from Simon's limited rationality, which refers to cognitive limitations and sub-ideal knowledge for rational choice: we claim that considering only active goals pertinent to the context is an ideal strategy.

(ii) qualitative relations, namely means-end links, and not only quantitative relations.

(b) In an AI perspective we think that the goal-strategy gives the agents more persistence, more predictability and more reliability. Agents will not pursue only their profit maximisation, being indifferent to one or the other goal, but will pursue specific goals they are committed to. One cannot commit itself to its profit: one should commit itself to a specific future intention.

5. A COMPUTER-SIMULATION STUDY OF PARTNERSHIP FORMATION: THE MICRODEP SYSTEM

On the tracks of DEPNET, a computational system, MICROdep (cf. Conte & Sichman 1995 and Conte et al. 1996), has been developed and employed to

(a) describe social agents as endowed with

(i) different goals and actions (*substantial* diversity), and

(ii) different principles of decision-making (*architectural* diversity), based on either utility-maximisation (aiming at maximising the overall utility independent of the values of the single goals) or goal-satisfaction (aiming at achieving goals with the highest values);

(b) calculate a general matrix including all the agents' preference lists of possible partners, as determined by pre-existing dependence relations;

(c) form real partnerhips thanks to successive prunings and updatings of this initial matrix, up to completion of the list.

In our simulations, agents are situated in a social market. The agents' values in a social market is determined neither from the chances they have to ask many people for help, nor from one's capacity to help others only, but from the intersection between these two factors. The more the agents which need one's help among those one is in need of, the higher the chances one has to find people with which it is both possible and convenient to interact. Consequently, the more likely one is to achieve one's goals. This intuition has been incorporated (cf. Conte & Castelfranchi, 1996) into the notion of *negotiation power*.

A single agent's negotiation power (negot-power for short) is a function of the intersection between the set of agents she *OR*-depends upon and the set of agents

CO-depending upon her. The larger the intersection and the higher the agent's negot-power. In order to easily compare the agents' negot-power within one social market, the intersection is weighted against the total agents' number -1. (However, if agents from different markets need to be compared, the formula should be modified).

5.1. The main hypthesis: the predictive role of the negotiation power

MICROPdep, which represents a development ofDEPNET, has been worked out in order to simulate the formation of partnerships in social markets. First, the dependence network for a given social market and each agent's negot-power are calculated; secondly, an algorithm for identifying each agent's list of preferred partners is applied; finally, an algorithm for extracting actual partnerships out of the global list ofpreferences is applied.

The central hypothesis underlying MICROdep is that the dependence network allows the formation ofpartnerships to be predicted. In order to test this hypothesis, we decided to check the predictivity of power negotiation. More specifically, in order for our hypothesis to be tested, each agent's negot-power in a given market should be checked against:

(a) the frequency ofthat agent's appearance in the others' preference lists, that is, the degree of her social utility: the higher the agent's negot-power, and the more frequently she should appear;

(b) the probability ofher appearance in the partnerships actually formed, that is, the extent to which the agent is able to form real partnerships: the higher her negot-power and the more chances she has to find partners;

(c) the rewarding potential of actual partnerships: the higher one's negot-power and the higher the outcomes one should get from the partnership entered.

5.2. What MICROdep does.

The system includes three main functionalities: (a) "reading" a social market, namely the agency which the user feeds the system with, and calculating both the dependence network corresponding to that market, and the single agents' negot-powers; (b) identifying the agents' preference lists; (c) calculating the actual partnerships.

Let us see each ofthese functions at some length.

5.2.1. The social market and the dependence network

MICRODEP reads the description of a given market, consisting of a given number of agents with their individual characteristics. To simplify matters, in the present version ofMICROdep, resources and plans are dropped; agents may differ in terms of *what they have*, that is, their actions and resources, and *what they want*, that is, their goals.

Thenafter, it computes whether any of the agents is autonomous for reaching its goals.

In our terms, an agent is socially autonomous when she has got all actions required to achieve her goals; an agent is socially dependent when it is lacking at least one action required by her goals and someone else has it. Within our perspective, an autonomous agent has a null negot-power, because she doesn't need to create partnerships: she calls herself "out of the game".

Table 1. External description of a 3-agent market

Market	Actions/Res.	Goals	Dependence
Agent 1	a	c	Y(2)
Agent 2	c	a	Y(1)
Agent 3	b	a	Y(1)

Agent 1 has the action/resource a, achieving the goals of both Agent 2 and Agent 3. But Agent 1 needs only Agent 2, whose action/resource fulfils Agent 1's goal. Therefore, Agents 1 and 2 should have the same negot-power.

MICRODEP computes now the NegotP of the three agents.

$$NegotP_{(1)} = 0.5,$$
$$NegotP_{(2)} = 0.5,$$
$$NegotP_{(3)} = 0.$$

5.2.2. *Heuristics for social choice*

As was said before, there are at least two possible "characters" or heuristics which agents may resort to for choosing their partners; these are two distinct criteria for producing each agent's preference list of possible partners, the *goal-* and the *gain-*oriented criteria. In the former, each agent orders her potential partners according to the relative goal-satisfaction she can get out of each of them and, but only as a second-order factor, to the costs she should sustain for those partnerships. Therefore, if $agent_i$ obtains from $agent_j$ the satisfaction of four goals, and from $agent_k$ the satisfaction of three goals, she will prefer $agent_j$ over $agent_k$ even though the payoff of the former partnership happens to be lower (suppose she must reward $agent_j$ with, say, three actions) than the payoff of the latter (which we may suppose to cost her, say, only one action). In the gain-oriented strategy, instead, the goal-satisfaction has no priority over the principle of utility. Therefore, in the above example, the preference order would be inverted, and $agent_k$ would be preferred over $agent_j$. With the strategy in question, we endeavoured to account for one important difference between rational agenthood and other architectures of intelligent systems, namely the difference between an abstract, undeterminate function of utility, usually attributed to rational agents, and the principle of the importance of concrete, specific goals, which is implemented in many intelligent natural systems, such as humans. According to the latter principle, agents do not act in order to maximise their utility, but rather to achieve the most important[5] of their goals, if more than one.

By definition, the gain-oriented strategy is more advantageous than the other from a strictest utilitarian perspective. Overtime, the agents following this strategy should sustain lower costs for entering partnerships than the agents following the alternative option. However, there could be individual values, other than the cost to benefit relation, which vary as a positive function of the agents applying the goal-oriented strategy, e.g. "personal" satisfaction, "well-being", "health", all notions which, in the context of simulation, have only a metaphorical savour, but which acquire a more explicit interest when applied to natural organisms and real populations. However, simulation allows these as well as other strategies of decision-making to be compared, and their relative roles to be identified and measured. The present study aimed at exploring alternative mechanisms of decision-making and observing their behaviors in the context ofpartnership formation.

5.2.3. MICROdep preference lists

Each agent has a preference list formed by all other agents, ordinally arranged, within the same market. In creating the agent's subjective preference list, either the goal- or the gain-oriented heuristics must be chosen, so that each list is formed either by considering how much each agent can receive by the others, or else, how many goals can be satisfied (independent ofhow much the agent pays for it). The agents one is not dependent upon have a null score on one's preference list. Completely autonomous agents have a null preference list.

The preference lists of the agents are given by the decreasing ordered series of those agents which they gradually depend upon. In the goal strategy, for example, if $agent_i$ depends on $agents_{k,\ l,\ m}$, that respectively fulfil two, four and three goals of $agent_i$, her preference list will be serially formed by $_{l,\ m}$, and $_k$. In the gain strategy, if $agent_i$ must return them respectively one, four and five actions, her preference list will be serially formed by $agents_{k,\ l,\ m}$.

5.2.4. MICROdep algorithm for partnership formation

MICROdep algorithm for partnership formation has been fully described in Conte et al., in press.[6]

MICROdep simulates the formation ofreal partnerships by producing a list of all possible exchange dyads, and by repeatedly pruning this list. In each subsequent pruning, the relative best choice are formed and removed from the market. The matrix is updated after each pruning, and this function is applied again and again until partnerships are possible (dependent agents exist and match one another).[7]

5.3. Findings

MICROdep was applied to

(a) test the hypotheses enunciated above (§ 4.1), and assess whether a dependency-based notion such negotiation power allows partnership formation to be predicted, and

(b) explore the difference between different strategies of decision-making in the formation of partnerships. In this study, two such strategies were examined, the goal-oriented and the gain-oriented strategy, the former being defined as a strategy aimed at achieving as many goals as possible independent of their costs, and the latter being aimed at maximising utility.

Three specific hypotheses were examined, namely whether the negot-power of each agent correlates with (a) the degree of her social utility (her appearance in the others' preference lists); (b) her probability of appearance in real partnerships, and (c) the values of these partnerships (p.v.). The higher the agent negot-power, and the higher each of these values was expected to be.

Simulations have been carried out with markets of different size. Three sets of 10 simulations have been run, respectively with 50, 30, and 10 agents. Results are shown in Tables 4 to 6. Each table shows the average values of the findings (correlations) for each set of simulations, that is, for each market size: Table 4 refers to findings in 10-agents markets; Table 5 to 30-agents markets, and Table 6 to 50-agents markets. In each table, the correlations between negot-power and four other values are shown:

(a) SU1, which refers to the agents' Social Utility type one, that is, to the frequency with which each agent appears in the list of *preference* of the other agents in the same market formed by applying the *goal*-oriented strategy;

(b) P.V.1, which refers to the values of the *partnerships* formed by following the *goal*-oriented strategy;

(c) SU2, which refers to the agents' Social Utility type two, that is, to the frequency with which each agent appears in the list of *preference* of the other agents in the same market formed by applying the *gain*-oriented strategy;

(d) P.V.2, which refers to the values of the *partnerships* formed by following the *gain*-oriented strategy.

As the tables show, all our *hypotheses* were *confirmed.*

(a) The agents' negotiation powers shows significant correlations with their social utility (see Tables 4-6, last rows), especially with the second type of social utility, based on the gain strategy; as will be shown below, the former type, based on the goal strategy, shows less stable correlations with any other dimension, including the negotiation power (see the first column of both Table 5 and 6). However, in larger markets, all correlations are found significant (see Table 6).

(b) The latter two hypotheses —namely whether the agents' negot-powers is significantly correlated with partnership formation, on one hand, and with the value ofpartnership, on the other— were also confirmed (see again the last rows ofall tables).

However, some remarks are necessary. First, the *findings obtained in our simulations vary with the market size.* In small markets, the predictivity of all the dimensions is considerably lower than in large and even middle-size markets (see Table 4).

The negotiation power is by far the most predictive dimension, but all correlations, although almost always significant, are lower than in larger markets. Furthermore, the correlations among the other dimensions are generally not significant, except that between the two types of social utility (p=.024), on one hand, and between the two types ofpartnership values, on the other (p=.000).

Table 4. Correlations in small-size markets

	SU1	PV1	SU2	PV2	Negot-power
SU1	10000	2745	8190	2509	3345
	p= ---	p=551	**p=024**	p=587	p=463
PV1	2745	10000	5601	9922	7751
	p=551	p= ---	p=191	**p=000**	**p=041**
SU2	8190	5601	10000	5957	7739
	p=024	p=191	p= ---	p=158	**p=041**
PV2	2509	9922	5957	10000	8316
	p=587	**p=000**	p=158	p= ---	**p=020**
Negot-power	3345	7751	7739	8316	10000
	p=463	**p=041**	**p=041**	**p=020**	p= ---

Marked correlations are significant at p < 05000

N=10 (Casewise deletion of missing data)

In middle size markets, all findings improve considerably, except the first type of social utility, which never shows signific*ant correlations (see Table 5).*

In large markets, (50 agents) not only the correlation between the negot-power and each of the remaining four dimensions are found significant (see Table 6; significant correlations are in bold), but also the correlations between any dimension and each ofthe others turned out to be significant.

In other words, in large markets, the agents' *negotiation power* is highly predictive of

(a) social utility in both the modalities of preference list formation that were implemented (the goal-oriented and the gain-oriented);

Table 5. Correlations in middle-size markets

	SU1	PV1	SU2	PV2	Negot-power
SU1	10000	-0628	-2776	-2619	-0543
	p= ---	p=720	p=106	p=129	p=757
PV1	-0628	10000	7224	7957	6164
	p=720	p= ---	**p=000**	**p=000**	**p=000**
SU2	-2776	7224	10000	8674	6703
	p=106	**p=000**	p= ---	**p=000**	**p=000**
PV2	-2619	7957	8674	10000	7277
	p=129	**p=000**	**p=000**	p= ---	**p=000**
Negot-power	-0543	6164	6703	7277	10000
	p=757	**p=000**	**p=000**	**p=000**	p= ---

Marked correlations are significant at p < 05000
N=30 (Casewise deletion of missing data).

Table 6. Correlations in large-size markets

	SU1	P.V.1	SU2	P.V.2	Negot-power
SU1	1.0000	.6099	.5814	.3152	.6595
	p= ---	**p=.000**	**p=.000**	**p=.031**	**p=.000**
P.V.1	.6099	1.0000	.6680	.3738	.7737
	p=.000	p= ---	**p=.000**	**p=.010**	**p=.000**
SU2	.5814	.6680	1.0000	.2892	.7618
	p=.000	**p=.000**	p= ---	**p=.049**	**p=.000**
P.V.2	.3152	.3738	.2892	1.0000	.3885
	p=.031	**p=.010**	**p=.049**	p= ---	**p=.007**
Negot-power	.6595	.7737	.7618	.3885	1.0000
	p=.000	**p=.000**	**p=.000**	**p=.007**	p= ---

Marked correlations are significant at p < .05000
N=50 (Casewise deletion of missing data)

(b) partnership values, again in both modalities.

To sum up, the larger the market, and the more likely a dependency-based notion such as negotiation power is to predict the formation of partnerships and their values. This can but strengthen the predictive value of the model considered, which proves more robust in larger and qualitatively rich distribution of individual characteristics. In particular, the scarcity of social resources (possible partners), which is a phenomenon obviously more frequent in smaller than large markets, may prevent even high negot-power agents from finding partners.

A second interesting consideration suggested by our findings concerns a *hierarchical order* among the dimensions considered. The negotiation power is by far the most predictive dimension.

Finally, some words should be spent about the *comparison between the two modalities of partnership formation*, the goal-oriented and the gain-oriented. Our findings show both similarities and differences between them. As far as the similarities are concerned, it should be noted that,

(a) in large markets, all dimensions in both modalities are significantly correlated,

(b) the negotiation power is predictive of all dimensions in both modalities (except the first type ofsocial utility) even in middle and small size markets,

(c) generally speaking, in all markets, the two modalities show significant correlations: at least in large size markets, agents which are mostly preferred in order to satisfy one's goals (first type of social utility), are also likely to be preferred in order to maximize one's utility (second type of social utility); analogously, partnership value in one modality is always significantly correlated with the alternative modality.

Therefore, the two strategies show quite similar behaviors, and seem to predict each other to a considerable extent. But our findings show also that social utility of the goal type, in small and middle markets, is not predicted by any of the other dimensions considered, including the negotiation power. This is not surprising, since agents which are socially useful to satisfy others' goals may be depending from them to a high degree (they may need much more than what they can give). For statistic reasons (higher variance), this phenomenon can be considered as more likely to occur in small than in large markets. With the social utility of the second type, instead, where the cost to benefit relation is necessarily considered, such a possibility is relatively less likely to occur: a useful agent, in terms of profit, is one whose "price", so to speak, must be lower than what she has to offer.

As was observed earlier in this paper, a more interesting comparison among these two modalities would imply, first, that different goal values be implemented; secondly, and more importantly, that the effects of these two strategies on distinct individual measures of adaptiveness —including rationality as well as, say, "health"— be observed and compared.

6. CONCLUSIONS AND FUTURE WORK

In this paper, a model for describing dependence relations among agents endowed with different actions and goals was applied to a computer-simulation study of partnership formation. This paper revolves around the idea that a double process of emergence should be described to account for partnership formation. More explicitly, social structures (e.g., dependence networks) are hypothesized to be derived from lower-level phenomena (individual properties), and, in turn, partnerships are expected to emerge from dependence relations.

In order to observe this double emergence process, a computational system, MICROdep, has been developed on the tracks of an existing tool, called DEPNET, and based on a model of social dependence. MICROdep was employed to (a) describe social markets, in which agents endowed with different goals and actions, and therefore with different negotiation powers, may depend on one another to achieve their goals; (b) calculate a general matrix including all the agents' preference lists of possible partners, as determined by pre-existing dependence relations; (c) form real partnerships thanks to successive prunings and updatings of this initial matrix, up to completion of the list.

The findings obtained by repeated simulation runs were recorded, in terms of how often each agent enters the other agents' preference lists; which agent forms real partnerships, and what is the value of the partnerships formed. These measures were finally correlated with the agents' negotiation powers. Findings were found to confirm the expectations: the agents' negotiation powers are highly correlated with almost all the measures obtained from the simulations, especially in large-size markets. This seems to show that partnerships can be expected on the grounds of dependence networks, and the latter can be predicted by describing a social aggregate in terms of individual properties. Social networks are not to be seen as necessarily *ad hoc* constructs, but as configurations determined by lower-level phenomena.

Our findings seem also to encourage the exploration through computer-simulation of different strategies of decision-making. So far, in our opinion, an essentially monarchic view of this phenomenon has prevailed in decision theory, namely the idea that it is rational for self-interested agents to maximise their utility, whatever the content of such utility may happen to be. In our study, we have endeavoured to show that self-interested agents can apply a somewhat different, although related, criterion for decision-making, namely goal-satisfaction. Of course, comparisons between these criteria, especially between their relative effects on the agents applying them, are still to be made. However, such a comparison should be made on a variety of measures, including but not reduced to those analytically consistent with the utility principle (accumulation, capitalisation, and the like). This is one of the objectives which MICROdep will be applied to in the future.

Other predictable developments of the system will include:

(a) the formation of multi-agent coalitions, rather than mere dyadic partnerships, in order to explore the emergence of spontaneous communities, informal organisations, etc.;

(b) the implementation of agents migrations from one market to another; this development actually implies

(i) a slightly different formal definition of negotiation power: agents from different social markets should compare with one another by estimating their own negotiation powers on different existing markets

(ii) the implementation of criteria for deciding upon migration; again, different criteria for such a decision could be identified and compared.

*Rosaria Conte**
CNR. Roma
Italy

NOTES

* This work has been realized with contribution of the ESPRIT Working Group on MODELling AGEncy.

[1] For example, the difference between short-term vs long-term rationality (think of one-shot and re-iterated games); the size of the social group; the resource typology (finite, even scarse; consumable vs self-replenishing, etc.).

[2]For the formal expression of our model, see Sichman et al. 1994.

[3]For the sake of brevity, we will ignore the case of multi-party dependence with regard to different goals, which, by the way, is but a multiple two-party dependence.

[4] Suppose x wants to have a baby. Given her preference relations in p_i, she will act as to increase the probabilities that she finally gets pregnant. However, she suddenly realises that given the financial market's current condition she should invest a considerable amount of money and start a new economic activity on some given field of production, since by doing so her overall utility—as is perceived by her—would increase to a higher extent than what would be the case if she only puts to execution her initial project (to have a baby): while having a baby scores high in her preference list p_i, it has none or negative consequence for other clusters of preferences p_j or p_z, for example make good money, have success, take a personal revenge against her former husband, etc. It might be the case that, once dropped her former plan (to have a baby) which was found incompatible with a more convenient course of action (to start a new entrepreneurial activity), x finds out that she is inheriting a large sum of money from her ex-husband who in the meantime has died. In the current situation, her overall utility is increased more by having a child, which is something x is still lacking, than by enlarging her already good incomes. Therefore, she is bound to dismantle or abandon her promising activity and retrieve motherhood.

This paradoxical example is used to illustrate how maximisation of utility may lead to deadlocks or stalemates in decision-making. Unlike what might be argued, these phenomena do not depend on unpredictable events and bounded information, but rather on the mechanism of decision-making itself: since what counts is the final utility obtained, it should be totally indifferent *how* that value is realised.

[5]Unfortunately, goal-importance is only partially expressed by the goal-oriented strategy as implemented in MICROdep at the current level of its development, since the importance of the goal is transformed into the number of goals. Future advances of the system are meant to implement goals with different values. However, even within the present restrictions, some comparison between the two strategies could be made.

[6] A matrix M formed by n rows, corresponding to the preference lists of the n agents in the market, is created. Each preference list is a vector containing at most n-1 non-empty items, that is, as many as the (n-1) agents in the market that a single agent can choose as partners for exchange. To the non-empty items included in the first column (that is, to the "first best" in each list of preference), a value (a.v.) equal to n-1 is associated; to the non-empty items of the second column, a a.v. equal to n-2 (second-best) is associated, and so on up to the non-empty items of the last column, whose a.v. is equal to 1. The empty items have an a.v. of 0 (zero). In Table 2, the matrix of preferences relative to a market including seven agents is shown.

TABLE 2. The matrix (7,6) of preferences relative to Market A, including seven agents

Agent A	C (6)	E (5)	G (4)	(0)	(0)	(0)
Agent B	A "	C "	E "	G (3)	"	"
Agent C	D "	B "	F "	A "	E (2)	G (1)
Agent D	A "	G "	E "	B "	(0)	(0)
Agent E	F "	G "	A "	C "	"	"
Agent F	G "	C "	B "	D "	A (2)	E (1)
Agent G	A "	C "	E "	B "	D "	F "

[7] Let us see how this is done. Each possible dyad is characterised by (a) a global value called p.v. (partnership value), equal to the sum of the agents' respective a.v.s, and (b) the absolute value of the difference between the agents' respective a.v.s. When either a.v. is null, the global value p.v. is also null, because possible dyads in which one of the partners is not included in the other's preference list are not considered. This is the situation of a Market's potential partnerships at time t_0. All the partnerships that will be actually formed will maintain the p.v. they have at time t_0. MICROdep updates the matrix M, and produces M^1 (see Table 3), where only agents whose a.v.s, if summed up, give a non-null value are considered (that is, agents which appear each in the other's list of preference). New a.v.s are now assigned to the items in the matrix.

TABLE 3. M^1 the matrix corresponding to Market A at time t_1

Agent A	C (6)	E (5)	G (4)	(0)	(0)	(0)
Agent B	C "	G "	(0)	"	"	"
Agent C	B "	F "	A (4)	E (3)	G (2)	"
Agent D	G "	(0)	(0)	(0)	(0)	(0)
Agent E	F "	G (5)	A (4)	C (3)	"	"
Agent F	G "	C "	E "	(0)	(0)	(0)
Agent G	A "	C "	E "	B (3)	D (2)	F (1)

In M^1, possible dyads are characterised by new p.v.s. Now, MICROdep :

(a) selects from M^1 the dyad(s) with the highest p.v. and the lowest |D a.v.|;
(b) removes from the matrix all dyads which have formed real partnerships.

If some dyads are left, a new matrix, M^2, is produced. In M^2, values are updated. Left dyads are picked, and the agents which they include are removed from the matrix, and values are updated again. This is recursively applied until no other possible partnership is left. Then, the computation stops.

REFERENCES

Boman, M., & Ekenberg, L., 1995: "Decision Making Agents withy Relatively Unbounded Rationality", in *DIMAS'95* - Proceedings of the First International Workshop on Decentralized Intelligent and Multiagent Systems - 22 Nov.1995, Krakov, Poland, 28-37.

Castelfranchi, C., Miceli, M., & Cesta, A., 1992: "Dependence relations among autonomous agents", in Y. Demazeau & E. Werner (eds.), *Decentralized AI - 3*, Elsevier, North-Holland.

Conte, R., 1996: "Foundations of rational interaction in cognitive agents: A computational approach", in W. Liebrand and D. Messick (eds), *Frontiers in social dilemmas*, Springer, Berlin.

Conte, R., & Castelfranchi, C., 1995: *Cognitive and Social Action*, UCL Press, London.

Conte, R., & Castelfranchi, C., 1996: Simulating interdependence networks. A two-way approach to the micro-macro link, in K.G. Troitzsch, U. Mueller, N. Gilbert, J. Doran (eds.), *Social Sience Microsimulation*, Heidelberg, Springer.

Conte, R., & Sichman, J., 1995: "DEPNET: How to benefit from social dependence", *Journal of Mathematical Sociology* ***20***(2-3) 161-177.

Conte, R., Veneziano, V., Castelfranchi, C., 1996: The computer simulation of partnership formation. *Computational and Mathematical Organisation Theory*, in press.

Gmytrasiewicz, P.J., 1995: "On Reasoning About Other Agents", in DIMAS'95 - *Proceedings of the First International Workshop on Decentralized Intelligent and Multiagent Systems* - 22 Nov.1995, Krakov, Poland (p.I/38-49).

Halpern, J.Y., and Moses, Y., 1984: "Knowledge and common knowledge in a distributed environment", in *ACM Conference on Principles of Distributed Computing*.

Hargreaves-Heap, S., 1993: "Post-modernity and new conceptions of rationality in economics", in B. Gerrard, (ed), *The economics of rationality*, Routledge, London.

Rao, A.S., & Georgeff, M.P., 1991: "Modelling rational agents within a BDI architecture", in J. Allen, R. Fikes, & Sandewall, E. (eds.), *Proceedings of the International Conference on Principles of Knowledge Representation and Reasoning*, Kaufmann, San Mateo,CA.

Rosenschein, J. S., & Zlotkin, G., 1994: *Rules of Encounter*, MIT Press, Cambridge, MA.

Sichman, J.S., Conte, R., Castelfranchi, C., & Demazeau, Y., 1994: "A social reasoning mechanism based on dependence networks", in A.G. Cohn (ed.), *Proceedings of the 11th. European Conference on Artificial Intelligence*, John Wiley & Sons, Baffins Lane, England, pp. 188-192.

Tuomela, R., 1996: *The importance of us*, Stanford University Press, Stanford.

Veneziano, V., Conte, R., Castelfranchi, C.: *MICROdep The formation of partnerships*, (in preparation).

Wellman, M.P., 1995: "Market-oriented programming: some early lessons", in S.H. Clearwater (ed.), *Market-Based Control: A Paradigm for Distrubuted Resource Allocation*, World Scientific.

BARBARA J. GROSZ

THE CONTEXTS OF COLLABORATION

1. INTRODUCTION

The term "context" originally applied to language and referred to the connections between words and sentences in a discourse. The word "context" comes from the Latin "contextere" meaning "to weave together." The Oxford English Dictionary (Simpson and Weiner, 1989) defines context as, "[t]he whole structure of a connected passage regarded in its bearing upon any of the parts which constitute it; the parts which constitute it; the parts which immediately precede or follow any particular passage or text and determine its meaning." Use of the word "context" has evolved so that it is now applied to actions more generally. Typically, context is used to refer to the various conditions in which an action is done or an event occurs. Context is seen to affect the ways in which actions are carried out and to influence the descriptions of events.

It has become clear to cognitive scientists and AI researchers that it is important to consider the ways in which thinking and acting, as well as communication, depend on the context in which they occur. Twin challenges arise for those interested in understanding the interactions of context with knowledge, mind, and society. First, the factors in an environment that affect descriptions of an event, choices of actions to undertake, or an individual's reasoning about actions and events must be identified. Second, the ways these various factors are woven together, to produce the context that affects interpretation or understanding of an event and the ways in which we process information about it, must be determined.

The interaction of context and language is two-way. Not only do contextual factors influence the meaning of linguistic expressions, but also language changes context. Among the clearest examples of the effect of language on context are uses of phrases such as "getting back to" that explicitly indicate a shift of context. Chapter and section titles in written works serve a similar purpose. Close examination of discourse reveals that almost every utterance has some effect on context. This characteristic also holds for the interactions between context and action. The context in which an action is performed affects the action and, where relevant, the interpretation of the action; in addition, the performance of the action changes the context.

Studies of discourse processing and discourse structure have established that the intentional setting of a discourse is an essential element of the context of communication (Grosz, 1977; Grosz and Sidner, 1986) and that an understanding of collaboration is essential to modeling the intentional context of discourse and its structure (Grosz and Sidner, 1990; Lochbaum, Grosz, and Sidner, 1990; Lochbaum,

K. Korta et al. (eds.), Cognition, Agency and Rationality, 175-187.

1995, 1998). As we have developed a theoretical framework for modeling collaboration (Grosz and Kraus, 1996), it has become evident that collaborative activity also has a complex set of contexts that must be taken into account.

This paper will first briefly examine the role of contexts in language processing and then turn to look more closely at the various aspects of context that affect collaborative activity. Although our primary focus will be on understanding the states of mind of the individuals who participate in a collaboration, we will be led nonetheless to ask about certain properties of the group, or society, as a whole. The next section briefly describes different aspects of context that affect understanding and production of language. The following section provides an overview of one model of collaboration to provide a setting in which to examine the roles of context in collaborative plans and activities. The final section discusses specific contextual factors and shows how the model that is described satisfies various criteria for an account of collaboration.

2. THE CONTEXTS OF LANGUAGE USE

Language use depends on contextual information to elaborate what has been literally said, removing ambiguity and further specifying content. In *The Lives of a Cell*, Lewis Thomas (1974, p. 94) wrote that "[a]mbiguity seems to be an essential, indispensable element for the transfer of information from one place to another by words, where matters of real importance are concerned." Contextual information helps determine meaning, decreasing or removing ambiguity. Because context can be called on to fill in missing information, speakers can communicate more succinctly. In their initial introduction to situation theory, Barwise and Perry (1983, p. 32) emphasize another facet of this property of language: "But what is important, after all, is the fact that expressions, whether simple or complex, can be recycled, can be used over and over again in different ways, places, and times and by different people, *to say different things*. This is what we mean by the *efficiency* of language."

This essential and powerful feature of natural languages has presented one of the major challenges to those who study language, whether from the perspective of understanding how people process language or from the perspective of constructing natural-language processing systems. Context is important at every level of analysis, from interpretation of the speech signal (e.g., in English, distinguishing "Mary" from "marry" or "merry", a job that may be more or less difficult depending on the dialect of the speaker) and identification of word meaning (e.g., determining whether "bank" is the side of a river or a financial institution) to composition of phrase or sentence meaning (e.g., determining whether the situation described by "Every student in the class watched a movie," is one in which the students all watched the same movie or one in which different movies were seen by different students) and the determination of utterance intention within a discourse (e.g., deciding whether the query "Do you know what time the train leaves?" is a request about mental state or a request for information). Furthermore, certain linguistic constructions, for instance definite descriptions and pronouns, are inherently context bound; interpretation of these expressions is impossible without taking context into account.

Significant research issues remain in determining appropriate ways to represent and reason about context so that the ambiguity that Thomas considers so powerful will increase rather than decrease the "efficiency of language" (in Barwise and Perry's sense) for computer systems. A range of contextual factors have been identified as relevant to natural-language processing, including the following:

- Global attentional state: the entities that are being talked about generally and the overall purposes ofthe discourse (Grosz, 1977; Grosz and Sidner, 1986);

- Local attentional state: memory for what has just been said and the entities and discourse purposes that are most salient as a result (Grosz and Sidner, 1986; Grosz, Joshi, and Weinstein, 1995);

- The (mutual) beliefs of the discourse participants: what is known to an individual participant or to all participants (Clark and Marshall, 1981; Cohen, 1981) as well as differences in belief among participants in a discourse (Pollack, 1990);

- The setting in which the discourse occurs, including both the location and the audience;

- The intentional context: the speaker's or writer's purposes, the "why's" for the discourse and its various constituents; the problem-solving situation, task, or goals ofthe speaker or writer.

It is this last element, the intentional context, that is of concern in this paper. Sidner and I (Grosz and Sidner, 1986) define three separate components of discourse structure: linguistic structure, attentional state, and intentional structure. We argue that intentional structure is the foundation on which the other two components rest. Relationships between discourse segments, which are the basic elements of linguistic structure, depend on relationships between the intentions corresponding to each segment. Changes ofattentional state are likewise guided by these intentional relationships.

Communication is a collaborative activity (Grosz and Sidner, 1990; Korta, 1995; Arrazola, 1996). As a result, theories and models of collaboration are essential to understanding and modeling intentional state and the intentional aspect ofdiscourse context (Grosz and Sidner, 1990; Lochbaum, Grosz, and Sidner, 1990). Lochbaum (1995, 1998) uses the SharedPlans formalization of collaboration, which is described briefly below, as the basis ofa computational model for recognizing the intentional structure ofdiscourse and shows how this model can be used in dialogue processing. Rich and Sidner (1994; also Sidner, 1997), building directly on Lochbaum's use ofSharedPlans, have constructed a collaborative graphical interface to a travel planning system. Both applications use the logical specification provided by SharedPlans to constrain utterance generation and interpretation.

There is also flow in the opposite direction: collaboration requires communication, and, like communication, collaboration is affected by a range of contextual factors. The remainder of this paper examines the contexts that affect an individual's reasoning about, and participation in, a collaborative activity.

3. THE SHAREDPLANS MODEL OF COLLABORATIVE ACTIVITY

Collaboration is one of several types of multi-participant activity. Its most distinguishing feature is that the participants in collaboration have a shared purpose or goal. Other kinds of multi-participant activity may entail taking into account various aspects of the behavior and needs of other agents. Collaboration—which is, literally, "working together" toward the accomplishment of shared goals—requires this and more. In particular it requires a range of commitments to the group activity—the shared purpose—and various related actions. Providing adequate treatment of these commitments is one of the major challenges to modeling collaborative activity.

The SharedPlans formalization of collaboration (Grosz and Sidner, 1990; Grosz and Kraus, 1996, 1998) defines the plans of a group in terms of the individual mental states of the group members. The formulation of collaborative plans in terms of mental state enables the following important characteristics of collaboration to be handled: the participants have different knowledge; they have different capabilities; they undertake different responsibilities related to their joint activity, which may both depend on and increase differences in knowledge; and plans for collaborative activity, like plans for individual activity, are often partial and evolve over time.

Although informed by studies of human behavior, the formalization is intended to be a normative account. It provides a specification for the construction of artificial agents (computer systems) that can collaborate with one another and with people. Hadad (1997) used the SharedPlans specification in designing a collaborative system for electronic commerce. Ortiz et al. (forthcoming) use it in the design of an interface to a distributed information system. In both these systems, the specification guided the design of the system architecture and constrained certain agent-planning processes (for example, to satisfy the axioms of intention). It also provided guidance about the information that the collaborating agents needed to establish for themselves and to communicate with one another.

The SharedPlans formalization is an individualistic account; there is no notion of irreducible joint intention. In this way it is similar to Bratman's account (Bratman, 1992) and in contrast to Searle's (1990). However, a major claim of our work in modeling collaboration is that collaborative activity is not merely the sum of individual plans (Grosz and Sidner, 1990; Grosz, 1996). Individuals do not simply plan or act on their own, hoping that their actions will merge appropriately to produce the desired behavior. Coordination, cooperation, and commitment to their joint activity are essential. The SharedPlans formalization provides these essential elements through a network of interlocking (individual) commitments to the joint activity, including commitments to the actions others take in behalf of this activity, and mutual beliefs of these commitments.

3.1. Intentions-that and Intentions-to

To represent the commitments participants have to a joint activity and to the actions of others within that joint activity has required the introduction of an additional intentional attitude. The SharedPlans formalization employs the attitude of *intending that* (Int.Th) a proposition hold in addition to the attitude of *intending to* (Int.To) do an action. Intending-that is an attitude that holds between an agent and a proposition, whereas intending-to is an attitude that holds between an agent and an action. In the formalization, Int.Th(G,prop,T_i,T_{prop},IC_{prop}) represents an agent G's intention at time T_i that a certain proposition T_{prop} hold in the intentional context IC_{prop} and Int.To(G, α,T_i,T_α,θ_α,IC_α) represents an agent G's intention at time T_i to do a certain action α at time T_α under the constraints θ_α and in the intentional context IC_α. As explained in a previous paper (Grosz and Kraus, 1996), the major differences between these two attitudes are in the kinds of means-ends reasoning they engender and in the assumptions each makes about an individual's ability to act, not merely in the types of objects toward which each is directed.

Intentions-to and intentions-that play different, but complementary and often interrelated, roles in collaborations. Some of their differences may be illustrated by the example of a professor, whom we will refer to as Professor Z, and a student, whom we will refer to as JD, writing a paper together.[1] Professor Z has agreed to write the introduction and JD has agreed to write the conclusions. Professor Z has an intention *to* write the introduction and an intention *that* JD be able to write the conclusion. JD has an intention *to* write the conclusion and intention *that* Professor Z be able to write the introduction. As a result of having an intention to write the introduction, Professor Z will perform certain other actions (for example, doing some background reading and writing certain prose) and will avoid making commitments to other activities that would conflict with this writing task. Her intention that JD be able to write the conclusion will lead her to avoid making requests that might cause conflicts with his intention to write the conclusion. Although her intention-that may lead her to undertake certain actions (for instance, advising JD on content or style), it will not engender the same direct means-ends reasoning as her intention to write the introduction.

3.2. Major Constituents of SharedPlans

The SharedPlans fomalization distinguishes between plans that are completely formed and those that are partial. Most often groups, like individuals, construct plans incrementally, and they interleave planning with acting. A group may determine only some of the steps to be taken in carrying out an activity; that is, they may have only a partial *recipe* for the action. Even when they have a complete recipe (know and agree on all the steps that need to be done), they may delay deciding who will do some steps. As a result, the SharedPlans formalization distinguishes between plans that are completely formed and those that are partial. In the formalization, FSP(P_α,GR,α,T_p,T_α,R_α,θ_α,IC_α) represents that the group GR has a full (i.e., complete) plan to do α at time T_α using the (complete) recipe R_α. The parameter θ_α denotes constraints on how α is done, for instance, constraints on the time or the location of the action. IC_α denotes the intentional context for the plan.[2] PSP(P_α,GR,α,T_p,T_α,θ_α,IC_α) represents that the group GR has a partial SharedPlan to do α at time T_α, under constraints θ_α and in context IC_α. Because we allow agents to have a partial SharedPlan before they have identified a recipe for doing their group activity,[3] there is no recipe parameter. Agents have a SharedPlan, represented as $SP(P_{\alpha},GR,\alpha,T_{p},T_{\alpha},\theta_{\alpha},IC_{\alpha})$, if either they have a full SharedPlan or they have a partial SharedPlan and a commitment to complete that plan.

SharedPlans, whether complete or partial, have four main elements. These elements are listed in Figure 1; formal definitions may be found in earlier papers (Grosz and Kraus, 1996, 1998). If a group of agents has a Full SharedPlan (FSP) to do a multi-agent action α, then they have (1) [individual] intentions that the group perform the action α; (2) mutual belief of a recipe for the action; (3) individual or group plans for the each of the constituent actions specified in the recipe, which we refer to as *subacts* (subsidiary actions); and, (4) [individual] intentions that each participant or subgroup succeed in these individual or group plans for the subacts.

Figure 1: Major constituents of a SharedPlan

For agents to have a SharedPlan to do a, they must have:

(1) individual intentions that the group perform α;
(2) mutual belief of a (partial) recipe for α;
(3) individual or group plans for the subacts in the (partial) recipe;
(4) intentions that the selected agents or subgroups succeed in doing their subacts (for all resolved subacts).

If the SharedPlan is partial, they must also have:

(5) individual intentions that the group complete the plan.

In full plans (FSP), the recipe [Clause (2)] is completely determined, some agent or subgroup has a complete plan for carrying out each subact [Clause (3)] and everyone in the group is committed to their being able to do so. In partial plans (PSP), the group may not have decided completely on the recipe for α or on who will do some of the subacts, or the plans for some of the subacts may be partial. Thus, in the case of partial plans, the mutual belief in Clause (2) may be merely that there is some recipe which the group will be able to find or figure out and then carry out. Also, some of the subacts may be "unresolved": before the plans stipulated in Clause (3) can be formed, the agents need to come to agreement about who will do an action, including forming commitments either to do the action or intentions that others will be able to do the action. Even in the case of partial plans, though, the intentions in Clause (1) must hold and those in Clause (4) must hold for all resolved actions. Partial plans have an additional element, Clause (5): the agents must have a commitment to complete the plan, that is, to identify the remaining actions in the recipe and to identify agents or subgroups to perform any as yet unresolved actions.

Certain planning and reasoning processes are essential to formulating group plans. Groups must have ways of expanding partial plans to more complete ones; we refer to these as Elaborate_Group actions. They must have ways of selecting recipes for actions, which could entail choosing from an existing library, combining ideas of different agents, reasoning from first principles, or some combination of these actions; we refer to these group procedures for selecting recipes as Select_Rec_GR. They must also have procedures for selecting agents (Select_Agent) or subgroups (Select_Subgroup) to take responsibility for performing subacts in the recipe. A description of these processes and their use in expanding partial plans to complete ones is provided in another paper (Grosz and Kraus, 1998).

Agents' intentions-that toward the successful performance of their collaborative activity [Clause (1) in Figure 1], toward their collaborators' ability to perform subacts [Clause (4)], and toward completion of their plan [Clause (5)] are key to achieving the collaboration needed for their joint actions to succeed. Axioms provided in the formalization (Grosz and Kraus, 1998) specify not only that agents avoid the adoption of intentions that conflict with the joint activity, but also that the subsidiary plans (individual and group) for doing the subsidiary actions in a recipe are compatible (or, "mesh" in Bratman's (1992) terms), and that agents endeavor to help each other in the performance of the group action and any subacts. These axioms indirectly give rise to communication actions, because communication is one way of correcting problems that arise in executing a plan or of helping another participant in the activity.

4. INTENTIONAL CONTEXTS FOR COLLABORATIVE ACTIVITY

One of the roles of intentional context in collaborative plans and activities is similar to its role in language processing. Agents interpret the actions or proposed actions of their partners against the background of their joint activity. In this way, intentional context constrains plan recognition. For instance, if one participant suggests doing an action, it is reasonable for other agents to consider how that action fits in the recipe for their SharedPlan or otherwise contributes to the activity. If Bill and Leslie are painting a house together, then Bill's offer to scrape the old paint off can be seen as an offer to do one of the constituent actions in the recipe for painting the house. Leslie's offer to buy a new ladder while at the store may be seen as helping make it possible to do some of the constituent actions in their plan. Lochbaum (1995, 1998) uses this kind of reasoning to determine how an utterance in a dialogue contributes to an evolving SharedPlan.

However, intentional context has additional roles in collaborative plans and activities. These roles relate to roles that Bratman (1987) argues intentions serve in plans, namely that they constrain the adoption of other intentions, focus means-ends reasoning, and direct replanning. When an agent adopts an intention to do some action, say β, in the context of a SharedPlan, then the SharedPlan provides the rationale for doing β. Bill scrapes off the old paint as part of a plan to paint the house with Leslie. If someone asks Bill to join a basketball game during the time he planned to scrape, then his consideration of whether to abandon his original intention (scraping) needs to take into account not only obligations to himself, as is the case in single-agent resource bounded reasoning (Bratman et al., 1988), but also to Leslie and their joint activity. The SharedPlan context may also influence an agent's choice of how to do an action and thus affect means-ends reasoning. Bill might use one recipe for scraping when working with Leslie and another when working alone. Finally, when agents incur problems during execution of a plan, the intentional context provided by the plan influences their replanning. For instance, the context might affect whether they decide to change recipes or look for different agents to carry out the action that is causing the problem. In all these cases, the intentional context of the SharedPlan provides key information by indicating the motivation for a particular intention, the other actions that are relevant to the group activity, and the commitments of the participants in the group activity.

4.1. Intentional Contexts in SharedPlans

In the SharedPlans formalization, an intentional context is associated with each of the intentional operators (Int.To, Int.Th) and with all plan meta-predicates (SP, PSP, FSP). In the formal notations for these operators and predicates given earlier, this context is represented by the last parameter (notated by IC). Table 1 states informally the intentional context for the major elements of a SharedPlan; a formal specification is given in Grosz and Kraus (1998).

Table 1: Intentional Contexts for Plans and Intentions

Intention or Plan	Intentional Context
G_k's intention to do β_i GR_k's partial SharedPlan to do β_k G_k's intention that GR_k be able to do β_k GR's plan to complete their partial plan for α GR's plan to find a recipe for α GR's plan to choose an agent to do β_i GR's plan to choose a subgroup to do β_k	GR's SharedPlan to do α, and the (partial) recipe being used in the plan
G_k's intention that the group identify all parameters for α G_k's intention that the group satisfy all constraints on doing α	GR's SharedPlan to do α

Legend: α is a group activity, GR is a group of agents, G_k is an agent in GR; GR_k is a subgroup of GR, β_i is an individual-agent subact and β_k is a multi-agent subact in recipe for α.

The first section of Table 1 gives the intentional contexts for various intentions and plans that are part of a SharedPlan to do α. If the action β_i, is part of the recipe for doing a, then the intentional context for an agent G_k's intention to do β_i, has two elements, the SharedPlan to do a and the possibly partial recipe associated with that plan, R_α^p. The SharedPlan for a provides information about the motivation for doing β_i, (namely, to do a), the other agents and actions involved in the plan, and the status of the plan. The recipe is an important element of the context when it becomes necessary to replan. For instance, if an agent discovers it cannot do β_k, it will need recipe information to determine whether a different recipe needs to be selected. The SharedPlan for a and the (partial) recipe R_α^p also provide the intentional context for a subgroup GR_k's partial SharedPlan to do a group action β_k that is part of the recipe for a, as well as for intentions, on the part of all group members, that GR_k be able to do β_k. The various planning and reasoning processes needed to expand the partial plan to a complete one —including Elaborate_Group, Select_Rec_GR, Select_Agent, and Select_Subgroup— also have both the SharedPlan and the recipe as their intentional context.

The intentions-that in the second section of Table 1 are used in SharedPlans to represent participants' commitments to meeting certain constraints on their actions. One of these is that agents must identify all of the parameters of the actions they have undertaken; this requirement is a knowledge precondition on the action (Lochbaum, 1995). The other deals with commitments to satisfying constraints on the action, for instance constraints on the location or time at which it is done. The intentional context for these intentions-that is only the SharedPlan. The recipe is not relevant, because the constraints are constraints on doing the action regardless of the method (or recipe) used.

As explained elsewhere (Grosz and Kraus, 1998), intentions *that* a proposition hold may lead to intentions *to* do actions that will either directly or indirectly help bring it about that the proposition holds. For instance, if Prof. Z intends that JD be able to write the conclusion this afternoon, but believes that JD can only successfully do this if someone else answers questions for new students, then Prof. Z might either intend to answer those questions (thus directly bringing about the proposition that JD can write) or intend to employ someone else to answer the questions (thus indirectly bringing about the proposition that JD can write). In either of these cases, the intention-that that spawned the intention-to provides the intentional context for it. In the indirect case, the intermediate action (e.g. JD's employment action) is also part of the intentional context.

4.2. Using Intentional Context to Solve a Puzzle

Because the intentional context provides the motivational context for a SharedPlan or an intention, it can be used to distinguish among different uses of the same action. For instance, it can be used to distinguish between the philanthropic actions of two types of business school graduates in the example Searle (1990) uses to argue against individualistic accounts of collaborative activity.[4] Searle describes two different classes of business school students. Figure 2 contains his descriptions of the classes, which we have labeled "Class 1" and "Class 2" for presentational purposes. Searle argues that Class 1 does not have "collective intentionality" whereas Class 2 does, and that, furthermore, an individualistic account cannot adequately treat this difference.

The combination of intentions-that and intentional context can be used to distinguish between these two classes, all within the individualistic approach of SharedPlans. Figure 3 contains the essential intentions that are part of the plans of the members of the different classes. Class 1 does not have a SharedPlan, but Class 2 does. The two classes have in common that each class member has an intention to "pursue his own self-interest", here represented as Int.To(S, Make_Money, . . .). However, the intentional contexts for these two intentions-to (IC1 and IC2 respectively) are different.

Figure 2: Searle's Business School "Counter-example" (Searle, 1990. pp. 404 - 405)

> Class 1:
>
> "Consider the following situation. Suppose a group of businessmen are all educated at a business school where they learn Adam Smith's theory of the hidden hand. Each comes to believe that he can best help humanity by pursuing his own selfish interest, and they each form a separate intention to this effect; that is, each has an intention he would express as "I intend to do my part toward helping humanity by pursuing my own selfish interest and not cooperating with anybody." Let us also suppose that the members of the group have a mutual belief to the effect that each intends to help humanity by pursuing his own selfish interests and that these intentions will probably be carried out with success. That is, we may suppose that each is so well indoctrinated by the business school that each believes that his selfish efforts will be successful in helping humanity."
>
> Class 2:
>
> ". . .the case where the business school graduates all get together on graduation day and [form] a pact to the effect that they will all go out together and help humanity by way of each pursuing his own selfish interests."

For Class 1, the intentional context is simply the individual class member's intention to help humanity (item (1b) in Figure 3). In contrast, for Class 2, the intentional context is the agent's intention that the whole class help humanity (2b). Furthermore, as part of their SharedPlan, the members of Class 2 each have this intention that the whole class help humanity, and they mutually believe that they have these intentions (2c). Intentional context IC2 and the intention-that in (2c) provide essential constraints on the agents' behavior. For example, as a result of the axioms of intention provided in Grosz and Kraus (1998), the members of Class 2 (but, crucially, not those of Class 1) are required to help each other in furtherance of their shared goal.

Figure 3: Relevant Intentions for Searle's Business School Graduates

> Class 1: For every student S in Class 1,
> (1a) Int.To (S, Make_Money, ..., IC1), where
> (1b) IC1: Int.To (S, Help_Hum, ...)
>
> Class 2: For every student S in Class 2,
> (2a) Int.To (S, Make_Money, ..., IC2), where
> (2b) IC2: Int.Th (S, Do(Class2, Help_Hum_Gr, ...), ...)
>
> Also, (2c) MB(Class2, Int.Th (S, Do (Class2, Help_Hum_Gr, ...), ...)) is part of Shared Plan

5. CONCLUSIONS

Communication and collaboration are interdependent. Communication is inherently a collaborative activity. People communicate with purpose, and the purposes underlying their communication provide structure to their discourse. Collaboration in turn requires communication. Both communication and collaboration are context dependent.

This paper briefly reviewed the roles of context in language processing. It then described a model of collaborative action that has been used both as a critical constituent of a computational model of dialogue participation and as the basis for constructing computer agents that collaborate with one another or with people. The paper discussed the intentional contexts important to collaborative activity. Finally, the paper demonstrated that the model of collaborative activity was able to distinguish a group collaborative activity from similar but non-collaborative actions of several individuals. The intentional contexts associated with various plan meta-predicates and intentional operators in the model are central to being able to make this distinction.

Barbara J. Grosz
Harvard University
U.S.A.

NOTES

* Partial support for the work reported here was provided by Grants No. IRI-9525915, IRI-9618848, and CDA 94-01024 from the National Science Foundation. The development of the formalization has been done jointly with Sarit Kraus. Luke Hunsbergeer and Charles Ortiz provided helpful comments on early drafts.

[1] We restrict the example to two participants only to simplify the presentation.

[2] Pa denotes the plan itself and Tp the time of the plan. These two parameters are not relevant to this paper and will not be mentioned further.

[3] The agents are required to believe they could determine a recipe and to have some plan for doing so. An explanation of this constraint may be found in an earlier paper (Grosz and Kraus, 1996).

[4] Grosz and Kraus (1998) make this claim, but do not show how the intentional context can be used to overcome Searle's objections.

REFERENCES

Arrazola, Xabier. 1996. "Collective Action: Logical Foundations for Interaction." Report No. ILCLI-96-FCSAI-1. Donostia - San Sebastian: ILCLI.

Barwise, Jon and John Perry. 1983. *Situations and Attitudes*. MIT Press. Cambridge, MA.

Bratman, Michael E. 1987. *Intention, Plans, and Practical Reason*. Harvard University Press, Cambridge, MA.Bratman, Michael E. 1992. "Shared Cooperative Activity." *The Philosophical Review*, 101. Pages 327-341.

Bratman, M., Israel, D., and Pollack, M. 1988. "Plans and Resource-Bounded Practical Reasoning." *Computational Intelligence*. Pages 349-355.

Clark, H. and Marshall, C. 1981. "Definite Reference and Mutual Knowledge." A.K. Joshi, B.L. Webber and I.A. Sag, editors. In *Elements of Discourse Understanding*. Cambridge University Press. Pages 10-63.

Cohen, Philip R. "The Need for Identification as a Planned Action." *In Proceedings of the International Joint Conference on Artificial Intelligence (IJCAI-81)*. 1981. Pages 31-36.

Grosz, Barbara J. 1977. The Representation and Use of Focus in a System for Understanding Dialogs. In *Proceedings of the Fifth International Joint Conference on Artificial Intelligence*. Cambridge, MA. Pages 67-76.

Grosz, Barbara J. "Collaborative Systems: 1994 AAAI Presidential Address." *AI Magazine*. Summer, 1996. Pages 67-85.

Grosz, Barbara, Aravind Joshi, and Scott Weinstein. 1995. "Centering: A Framework for Modeling the Local Coherence of Discourse." *Computational Linguistics*, 21(2):203--225.

Grosz, Barbara J. and Sarit Kraus. 1996. "Collaborative Plans for Complex Group Action." *Artificial Intelligence*, (86):269--357.

Grosz, Barbara J. and Sarit Kraus. 1998. "The Evolution of SharedPlans." In A. Rao and M. Wooldridge, editors, *Foundations and Theories of Rational Agencies*. To appear.

Grosz, Barbara and Candace Sidner. 1986. Attention, intentions, and the structure of discourse. *Computational Linguistics*, 12(3):175--204.

Grosz, Barbara and Candace Sidner. 1990. Plans for discourse. In P. Cohen, J. Morgan, and M. Pollack, editors, *Intentions in Communication*. Bradford Books/MIT Press, Cambridge, MA, pages 417--444.

Hadad, Meirav. 1997. "Cooperation Among Agents." Master's Thesis, Department of Computer Science at Bar-Ilan University, Ramat-Gan, Israel.

Korta, Kepa. 1995. "Mental States in Conversation." Report No. ILCLI-95-LIC-2. Donostia - San Sebastian: ILCLI.

Lochbaum, Karen E. 1995. "The Use of Knowledge Preconditions in Language Processing." In Chris S. Mellish, editor, *Proceedings of the International Joint Conference on Artificial Intelligence* (IJCAI-95), volume 2, pages 1260--1266, San Mateo, CA. Morgan Kaufmann Publishers, Inc.

Lochbaum, Karen E. 1998. "A Collaborative Model of Intentional Structure." *Computational Linguistics*. Forthcoming.

Lochbaum, Karen, Barbara Grosz, and Candace Sidner. 1990. "Models of Plans to Support Communication: An Initial Report." In *Proceedings of the 8th National Conference on Artificial Intelligence* (AAAI-90), pages 485--490, Cambridge, MA. MIT Press.

Ortiz, C., Grosz, B., and Scales, N. 1998. "Interpreting information requests in context: A collaborative web interface for distance learning." Forthcoming.

Pollack, Martha E. 1990. "Plans as Complex Mental Attitudes." In P.N. Cohen, J.L. Morgan, and M.E. Pollack, editors, *Intentions in Communication*. Bradford Books, MIT Press. Pages 77-103.

Rich, C. and Sidner, C. 1996. "Adding a Collaborative Agent to Direct-Manipulation." In *Proceedings: 9th ACM Symposium on User Interface Software and Technology*. Seattle, WA.

Searle, John R. 1990. "Collective Intentions and Actions." In *Intentions in Communication*. The MIT Press. Pages 401-415.

Sidner, Candace L. 1998. "Building a Collaborative Interface Agent." In Xabier Arrazola, Kepa Korta, and Francis Jeffry Pelletier, eds, *Discourse, Interaction, and Communication*. Kluwer Publishing.

Simpson, J. A. and E.S.C. Weiner, editors. 1989. *The Oxford English Dictionary*. Oxford: Clarendon Press, 2nd edition.

Thomas, Lewis. 1974. *The Lives of a Cell: Notes of a Biology Watcher*. Viking Press, New York.

SUBJECT INDEX

A

action, x, 33, 36, 116n, 125, 128-131, 135-140, 145, 146, 153, 154, 157-160, 162, 164, 165, 175, 178-180, 182, 183
- cooperative action, 140, 146
- individual action, 151
- joint action, 181
- social action, 145, 147, 151, 153-157
- *see also* interaction
- team action, 140

agency, vii, x

agent, x, 35, 125-132, 135-142, 145-147, 151, 171, 172n, 178-182, 184, 185, 186n
- autonomous agent, x, 135, 136, 139, 140, 145, 146, 151, 154, 164, 165
- cooperative agent, 140, 143, 155
- individual agent, 135-137, 139, 147, 156
- social agent, 146, 157, 158, 162, 164

agreement, 5, 99, 180

argumentative force, 106

argumentative theory, 113, 117

argumentative presupposition, *see* presupposition

artificial intelligence, vii, x, 38, 123-125, 127, 135, 151, 162, 175

distributed artificial intelligence, x, 151

autonomous agent, *see* agent

B

base-rate fallacy, *see* fallacy

BDI logics, *see* logic

belief, 1-4, 14, 33, 37, 55, 83, 88, 100, 101, 111, 124, 126, 129-132, 135, 153, 154, 177
- belief acquisition, 4
- belief ascription, 97
- belief fixation, 36
- belief formation, 2, 4
- belief management, 2
- belief sentences/statements, 82, 87
- disbelief, 2
- mutual belief, 177, 178, 180, 184

C

chance, 13-15, 26, 28, 41, 42, 142, 162, 163

Chomskian module, *see* module

code-breaking, 9

cognition, vii, x, 9, 36, 37, 57, 64

cognitive architecture, 35, 36, 47

cognitive capacity, 32

cognitive mecanism, 27-29, 37, 38, 41, 42, 47, 48, 64, 76

cognitive processes, 34, 64

cognitive science, vii, 9, 15, 18, 21, 22, 27-30, 32, 53-55, 123

cognitive structure, 35

cognitive task, 18, 32

collaboration, x, 175-179, 181, 185

communication, 18, 123, 126, 175, 177, 178, 181, 185

competence, vii, 1, 21, 22, 27-30, 32, 53-55

compositional meaning theories, *see* meaning

compositional semantics, *see* semantics

computation, 32-35, 37, 135, 147, 164, 173n, 177

computational module, *see* module

conditional, ix, 1, 2, 16, 63, 65-70, 73-75, 107, 110
conditional rule, 3, 18
counterfactual conditionals, ix, 65-69, 73-75
factual conditionals, ix, 65, 67-69, 73-75
mental models theory of conditionals, ix, 65
subjunctive conditionals, 70, 72, 73
theory of conditionals, ix, 65, 67, 75
conjunction, viii, 2, 14, 15, 84
conjunction fallacy, *see* fallacy
conjunction rule, 1, 15, 54
context, vii, ix, x, 5, 11, 14, 16, 17, 48, 49, 88, 93-95, 108, 109, 111, 112, 115, 116, 118, 125, 128, 132, 140, 146, 157, 161, 162, 165, 175-177, 179, 181-183, 185
context dependence, 93, 118, 185
context sensitive, 94-97
intentional context, x, 177, 179, 181-185, 186n
social context, 135, 147
contextualism, ix, 93-96, 101, 102, 103n
contrapositive reasoning, *see* reasoning
conversion, 42
illicit conversion, *see* fallacy
cooperation, 125, 127, 153-157, 178
counterfactualisable probability, *see* probability
counterfactuals, ix, *see* conditional

D
Darwinian module, *see* module
decision, 2, 21, 24, 25, 31, 128, 129, 131, 140, 143, 144, 151, 153, 161
collective decision making, 153
cooperative decision making, 140, 141, 143, 144
decision making, x, 135-142, 146, 147, 156, 161, 165, 166, 170, 171
decision theory, x, 41, 151, 152, 157, 160-170
individual decision, 5, 153
principle of decision making, 139, 162
social decision, x, 135, 137, 139, 142, 146, 147
deducibility, 10, 15-18
undisambiguated deducibility, *see* fallacy
deduction, 65

E
entailment, 107, 108, 112, 113, 118, 120n
pragmatic entailment, 107, 111, 113, 116
evolutionary psychology, *see* psychology

F
fallacy, 11-13, 15, 16, 18
base-rate fallacy, 11, 12
conjunction fallacy, 23, 29-31, 43, 44
fallacy of disambiguated probability, 12-14, 18
fallacy of illicit conversion, 16
fallacy of undisambiguated deducibility, viii, 16-18
gambler's fallacy, 13
focus, 1, 2, 105-107, 110, 113
focal particles, ix, 105, 106, 113
focal presupposition, *see* presupposition

G
gambler's fallacy, *see* fallacy
game theory, x, 151, 153
goal, 64, 124-126, 128, 130-132, 137, 140, 146, 151-161, 164-166, 160-170, 171n, 177, 178
common goal, 156
goal operator, 130-132

shared goal, 178, 184
social goal, 153, 157

H
helpfulness, 137, 139-144, 146
heuristic, 21, 30, 41, 52-54, 161, 164, 165
heuristic rule, 31
hidden relativity, *see* relativity

I
implicature, 16, 107, 110, 111, 119n
Gricean implicature, 10
implicit relativity, *see* relativity
indicative, 65, 73, 74, 82, 89
inference, ix, 4, 10, 21, 28, 36, 42, 55, 58n, 63-65, 69, 71-75, 79, 80, 87
counterfactual inference, 74
statistical inference, 19n
intention, ix, 16, 17, 32, 88, 98-101, 103n, 124, 126, 129-131, 152, 162, 176-185, 186n
communicative intention, ix, 97, 100
intention-that, x, 178-180, 183, 184
intention-to, x, 128, 131, 178, 179, 183, 184
intentional context, *see* context
intentional operator, x, 128-131, 182
intentional state, 126, 177, 185
joint intention, 178
meaning-intention problem, 93, 97, 100
interaction, x, 125, 128, 135, 136, 145, 146, 151, 156, 157, 161, 175
interest, 16
self-interest, 137, 139, 144, 155, 156, 179
social self-interest, 139, 141, 144
irrationality, ix, 2, 3, 6, 12, 15, 18, 22, 58n

L
language, 9, 27-31, 36, 54, 64, 69, 75, 80, 81, 85, 89, 90, 96, 97, 101, 102, 103n, 107, 113, 117-119, 127, 130, 132, 175-177, 181, 185
artificial language, 31
language game, 89
natural language, vii, ix, 31, 32, 89-91, 110, 176, 177
ordinary language, 17
philosophy of language, 79, 95, 96, 101, 102
logic, 1-3, 6, 17, 19n, 29, 58n, 116, 123, 126-128, 130
BDI logics, 126, 129-132
formal logic, 9
mental logic, 69
propositional logic/calculus, viii, 1, 17

M
meaning, ix, 11, 15, 79, 80, 82-84, 90, 91, 100, 107, 110, 120n, 124, 126, 175, 176
compositional theory of meaning, ix, 89, 90
meaning-intention problem, *see* intention
theory of meaning, ix
mental architecture, 22
mental logic, *see* logic
mental mechanism, 23, 45
mental model, 64
mental models theory of conditionals, *see* conditionals
mental module, *see* module
mental representation, 32, 33, 64, 69, 74, 76, 91, 153
mental state, x, 57, 79, 100, 101, 118, 123-130, 158, 176,178
mental structure, 32-34, 123
Mentalese, 90, 91
modularity, 32, 34-36, 125
massive modularity, 36-37, 39
Massive Modularity Hypothesis MMH, 36-38, 52-55, 57

module, 21, 22, 31-35, 37-40, 42, 47, 52, 55-57, 127
  Chomskian module, 32-36, 39, 54
  Darwinian module, viii, 34-36, 39, 41, 45, 47, 51-57
  mental module, 22, 32, 40, 47, 55
modus ponens, viii, 1, 3, 69, 71, 74
modus tollendo tollens, 16, 69, 71, 73, 74
multi-agent systems, vii, x, 125, 135-137, 140, 141, 145, 146, 153, 156, 157
mutual belief, *see* belief

N
natural selection, 21, 31, 35, 37-40, 53, 54, 56

P
paradox, 7, 8, 11, 91n, 96, 97
  skeptical paradox, 96, 97
  taxi-cabs paradox, 11
performance, x, 1, 19n, 114, 136, 142, 145, 146, 175, 181
plan, 64, 130, 156-160, 171n, 178-185, 186n
  collaborative plan, 176, 178, 181
  common plan, 156
  SharedPlan, x, 177-185
pleonastic concepts, ix, 84, 85, 87-90
pleonastic Fregeanism, ix, 89
pleonastic propositions, *see* propositions
posterior probability, *see* probability
pragmatic, vii, viii, ix, 105, 107, 112, 116, 118, 119n
  pragmatic entailment, *see* entailment
  pragmatic goals, 56
  pragmatic reasoning schemas, 51, 52
presupposition, 65-67, 105, 107-109, 116, 118, 119n
  argumentative presupposition, ix, 106, 107, 118, 119
  focal presupposition, 105-107, 113, 114, 117, 118
  pragmatic presupposition, 111
principle of social rationality, *see* rationality
probability, viii, 1-3, 10-16, 18, 19n, 24-26, 29, 30, 41, 42, 44, 45, 52, 53, 58n, 79, 137, 160, 163, 166, 171n
  posterior probability, 12
  prior probability, 11, 12
  probability theory, viii, 1, 9
  statistical probability, 1
  undisambiguated probability, *see* fallacy
proposition, ix, 1-4, 7, 15, 16, 67, 79-90, 93, 95-101, 111, 112, 128, 129, 179, 183
  categorical proposition, 67, 75
  Fregean theory of propositions, 81-83, 88
  pleonastic propositions, ix, 87-90
propositional attitudes, 81
propositional logic/calculus, *see* logic
psychology, vii, ix, 18n, 22, 32, 37, 53, 58n, 63, 157
  developmental psychology, 32
  evolutionary psychology, viii, 22, 23, 31-36, 39, 40, 42, 45
  experimental psychology, 9, 18n
  folk psychology, 57, 124

R
rationality, vii, viii, x, 1-5, 7, 10, 21, 52, 57, 69, 126, 130, 131, 139, 151, 152, 156, 157, 160, 162, 169
  economic rationality, x
  epistemic rationality, 2
  failure of rationality, 2, 4
  human rationality, viii, 2, 3, 6, 7, 22, 52, 57, 58n
  perfect rationality, 4
  practical rationality, 2
  principle of social rationality, 139, 146

rationality paradigm, x
social rationality, x, 147
strategic rationality, x, 151
reasoning, vii, ix, 4, 9, 10, 17, 18, 19n, 21, 21, 26, 27, 29-34, 36, 37, 40, 41, 44, 51-58n, 64, 65, 76, 117, 118, 126, 128, 130, 175, 177, 179-182
bayesian reasoning, 43, 45, 55
contrapositive reasoning, ix, 17
counterfactual reasoning, ix
deductive reasoning, viii
deontic resaoning, 51, 52
fallacious reasoning, viii, 10
inductive reasoning, 41, 42
probabilistic reasoning, 9, 15
rules of reasoning, 1
reference, 82, 86, 89, 94, 98, 99
reference class, 11, 14
relativity, 95, 99, 100
hidden relativity, ix, 95, 98, 100-102, 103n
implicit relativity, 95, 97, 98
responsibility, 178, 180
social responsibility, x, 135

S
self interest, *see* interest
semantic, vii, viii, ix, 14, 34, 69, 79, 93-97, 102, 125, 129-132, 105, 107, 113, 119, 120n
compositional semantics, vii, ix, 89-91
semantic value, 82, 83, 88
truth-functional semantics, 65
SharedPlan, *see* plan
skeptical paradox, *see* paradox
social responsibility, *see* responsibility
statistics, 24, 41, 42, 169
statistical probability, *see* probability
subjunctive, 1, 65, 74
subjunctive conditional, *see* conditional

T
thinking, 2, 13, 19n, 37, 47, 58n, 63, 64, 96, 97, 175
counterfactual thinking, 63, 64

V
variance, 12, 13, 169

NAME INDEX

A
Anscombre, J.C., 107, 112, 113, 117, 120
Arrazola, X., xi, 177, 186
Atxa, E., xi

B
Bach, K., 58n
Bar-Hillel, M., 19n, 58n
Barber, A., 91n
Barkow, J., 40, 59
Baron, J., 21, 59
Baron-Cohen, S., 34, 40, 59
Barwise, J., 176, 177, 186
Bayes, T., 11, 12, 42
Beach, L.R., 19n, 58n
Beckett, S., 63
Bishop, M., 58n
Boden, M., 58n
Bolinger, D., 119n, 120
Boman, M., 173
Borgida, E., 18n, 58n
Bower, G., 44, 60
Braine, M.D.S., 64, 76
Bratman, M.E., 126, 130-132, 178, 181, 186
Brenneman, K., 40, 60
Brooks, R.A., 127, 132
Browne, D., 58n
Byrne, R.M.J, vii, ix, 22, 63-65, 67-71, 73, 74, 75, 76, 77

C
Campbell, J. 76n
Campos, J.R., 138, 139, 148
Cara, F., 52, 61
Carey, S., 32, 40, 59
Carnap, R., 19n, 58n, 120
Carruthers, P., 59
Casscells, 26, 42, 43, 54, 59
Castelfranchi, C., 140, 145, 147, 151, 154, 155, 157-159, 162, 173
Caverni, J.P., 64, 77
Cesta, A., 146, 147
Chapman, J., 45, 59
Chapman, L., 45, 59
Chater, N., 52, 61
Cheng, P., 51, 59, 61, 64, 76
Chomsky, N., 27-30, 32, 33, 59
Clark, H.H., 177, 186
Clinton, H., 73
Cohen, L.J., vii, viii, 9, 18n, 19n, 58n, 59
Cohen, P.R., 126, 130-132, 177, 186
Cohen, S., 103
Conte, R., vii, x, 151, 154, 155, 157-159, 162, 165, 173
Copeland, J., 58n
Corkill, D.D., 136, 148
Cosmides, L., 35-49, 51, 53, 59, 61, 64, 76
Cox, J., 46, 60
Cox, M., 124, 127, 133
Cullicover, P.W., 119n, 120
Cummins, D., 52, 59

D
Daly, M., 40, 62
Davidson, D., 19n, 58n
Davis, R., 136, 147
Dawes, R., 21, 59
Demonte, V., 119n
Dennett, D., 124, 125, 133
DeRose, K., 103n, 104
Dibble, E., 66, 77
Downes, S., 58n
Doyle, J., 137, 147
Doyle, R., 63
Dretske, F., 103n, 104
Drogul, A., 136, 148
Ducrot, O., 107, 112, 113, 117, 120

E
Egan, M.F., 58n

Ekenberg, L, 173
Erman, L.D., 136, 147
Evans, J.St.B., 22, 61, 71

F
Fauconnier, G., 107, 112, 120
Ferber, J., 136, 148
Fiedler, K., 43, 44, 59
Fillenbaum, S., 66, 76
Fischer, K., 136, 148
Fischoff, B., 27, 30, 61
Fodor, J., 33, 34, 36, 59, 60, 91n
Foley, R., 58n
Foss, B., 19n, 58n
Franklin, S., 135, 148
Fraser, B., 108, 120
Freed, M., 124, 127, 133
Frege, G., 79, 85, 88, 89

G
Gallistel, C., 60
García Murga, F., vii-ix, 105-107, 111, 113, 116, 119n, 120, 121
Gardner, H., 60
Garfield, L., 34, 60
Gelman, S., 40, 60
Georgeff, M.P., 126, 131-133, 173
Gigerenzer, G., 10, 18n, 40, 41, 44, 45, 48, 49, 55, 58n, 60
Gillovich, T., 45, 60, 64, 76
Ginsberg, M.L., 64, 76
Girotto, V., 52, 61, 64, 65, 76, 77
Gluck, M., 44, 60
Gmytrasiewicz, P.J., 151-153, 156, 173
Gödel, K., 3, 129
Goldblatt, R., 126, 133
Goldman, A., 60, 103n, 104
Gould, S.J., 30, 31, 45, 52
Graesser, A., 135, 148
Graham, M., 136, 148
Grayboys, T., 59, 60
Grice, H.P., 19n, 58n, 103n, 104, 119n, 120n, 121
Griffiths, P., 36, 60
Griggs, R., 46, 60
Grosz, B.J., vii, x, 175-184, 186, 187

H
Hacking, I., 19n, 58n
Hadad, M., 178, 186
Halliday, M.A.K., 119n, 121
Halpern, J.Y., 126, 133, 153, 173
Handley, S.J., 69, 76
Hargreaves-Heap, S., 152, 160, 173
Harman, G., 19n, 58n
Harris, P.L., 19n, 58n
Haugeland, J., 33, 60
Heaney, S., 63
Hertwig, R., 44, 60
Hintikka, J., 126, 133
Hirschfeld, L., 60
Hoffrage, U., 41, 44, 60
Hofstadter, D.R., 64, 76
Hofweber, T., vii, ix, 93, 103, 104
Hogg, L.M., 147, 148
Holmes, S., 66
Holyoak, K., 51, 59, 64, 76
Horn, L., 106-109, 117, 121
Horty, J., 124, 127, 133
Hug, K., 40, 48, 49, 60
Hunsbergeer, L., 185n
Hutchins, E., 40, 60

J
Jackendoff, R., ix, 36, 60, 114, 119n, 120n, 121
Janca, P.C., 135, 148
Jennings, N.R., vii, x, 125, 133, 135, 138-140, 147, 148
Johnson-Laird, P.N., 19n, 58n, 63-70, 73, 75-77
Johnston, M., 86, 91n
Jones, C.J., 19n, 58n
Joshi, A.K., 177, 186

K
Kaelbling, L.P., 127, 133
Kahneman, D., 18n, 19n, 21, 23-25, 30, 43, 44, 53, 58n, 60, 62, 64, 77
Kalenka, S., vii, x, 135, 139, 140, 148

Kalokerinos, A., 120n, 121
Karmiloff-Smith, A., 60
Karttunen, L. 107, 110, 111, 121
Kay, P., 107, 112, 113, 120n, 121
Keane, M.T., 64, 77
de Keijzer, M. xi
Kempson, R., 108, 121
Kent, C., 81-83, 87-89
Ketchpel, S., 140, 148
Kleinbölting, H., 44, 60
König, E., 114, 121
Korta, K., xi, 177, 186
Kraus, S., 140, 148, 176, 178-185, 186
Kripke, S., 126, 129-133

L
Landman, J., 64, 77
Lane, L., 81
Legrenzi, M., 64, 77
Legrenzi, P., 64, 77
Lehman, D., 61
Lehrer, K., xi
Lempert, R., 61
LePore, E., 58n
Leslie, A., 40, 61
Lesser, V.D., 136, 147, 148
Levesque, H.J., 126, 130-132
Lewis, D., 65, 66, 73, 77, 103n, 104
Lichtenstein, S., 27, 30, 61
Lochbaum, K.E., 175, 177, 181, 183, 186, 187

M
Manktelov, K., 52, 61
Marr, D., 34
Marsh, S. 146, 148
Marshall, C., 177, 186
Mazzocco, A., 65, 76
McCarthy, J., 124, 125, 127-129, 133
McLaughlin, B., 58n
McNamara, J., 120n, 121
Medvec, V.H., 64, 76
Mendizabal, E., xi
Miller, G., 120n, 121
Miller, D., 64, 77
Miyamoto, J.M., 66, 77
Mohammed, 26
Moore, R.C., 130
Moses, Y., 126, 133, 153, 173

N
Napoleon, 26
Neander, K., 61
Nesbitt, R.E, 18n, 58n
Newell, A., 137, 139, 148
Newstead, S., 22, 61, 71
Nisbett, R., 21, 26, 51, 59, 61
Norvig, P., 125, 133

O
Oaksford, M., 52, 61
O'Brien, D.P., 64, 76
Oliver, L., 51, 59
Ortiz, C., 178, 185n, 187
Over, D., 52, 61
Overgaard, L., 136, 148

P
Peacocke, C., 81, 85
Perry, J., 103n, 104, 176, 177, 186
Peters, S., 107, 110, 111, 121
Peterson, C.R., 19n, 58n
Phillips, L., 27, 61
Piatelli-Palmarini, M., 21, 61
Pinker, S., 36, 37, 61
Plantinga, A., 61
Pollack, M.E., 177, 187
Pollock, J.L., 65, 77
Pylyshyn, Z., 33, 61

Q
Queen Victoria, 26
Quine, W.V.O, 79

R
Ramsey, W., 65, 77
Rao, A.S., 126, 131-133, 173
Reagan, R., 23, 24
Rich, C., 177, 187
Rijgersberg, R., xi
Rips, L.J., 64, 77
Rochemont, M., 119n
Roese, N.J., 64, 77
Rogers, A., 121

Rosenschein, J.S., 173
Rosenschein, S.J., 127, 133
Ross, J., 21, 26, 61
Russell, B., 85, 91n
Russell, S., 125, 133, 137, 148

S
Sales, T., vii, x, 123
Samuels, R., vii, viii, 21, 57, 61
Sánchez de Zavala, V., vii, 117, 121
Satan, 87, 88
Schaeken, W., 68, 70, 77
Schiffer, S., vii, ix, 79, 91n, 96-98, 101, 102, 103n, 104
Schmerling, S., 106, 121
Schoenberger, A., 59
Scholl, B., 58n
Scribner, S, 19n, 58n
Searle, J.R., x, 124, 125, 178, 183-185, 186n, 187
Segal, G., 31, 33, 34, 61
Sen, S., 146, 148
Shapiro, D., 19n, 58n
Shehory, A., 140, 148
Shoham, Y., 124, 126, 127, 129-131, 133
Sichman, J.S., 146, 148, 155, 157, 158, 162, 171n, 173
Sidner, C.L., 175, 177, 178, 186, 187
Simon, H., 162
Simpson, J.A., 175, 187
Singh, M.P., 127, 133
Slovic, P., 21, 30, 60, 61
Smith, P., 59
Smith, R..G., 136, 147
Sosa, E., vii, viii, xi, 1, 58n
Spelke, E., 32, 40, 59
Sperber, D., 36, 52, 56, 61, 119n, 121
Stalin, J., 66
Stalnaker, R.B., 65, 73, 77, 91n, 111, 121
Steels, L., 127, 133, 136, 148
Stein, E., 8n, 61, 133
Stich, S., vii, viii, 21, 40, 61
Superman, 81-83, 87-89
Sutherland, S., 21, 61

T
Tancredi, C., 106, 121
Tasso, A., vii, ix, 63-65, 74, 76
Thomas, L., 176, 177, 187
Thompson, V., vii, ix, 63, 65, 74, 75, 77
Tooby, J., 35-49, 51, 53, 59, 61
Tremoulet, P.D., vii, viii, 21
Trivers, R., 47, 61
Truman, H., 87
Tuomela, R., 174
Tverski, A., 18n, 19n, 21, 23-25, 30, 43-45, 53, 58n, 60, 62

V
Vallone, B., 45, 60
Veneziano, V., 174
Voogd, J., xi

W
Wason, P.C., 17, 19n, 22, 23, 46, 48, 58n, 73
Watson, Dr., 66
Wavish, P., 136, 148
Wefald, E., 137, 148
Weiner, E.S.C., 175, 187
Weinstein, S., 177, 186
Wellman, M.P., 137, 149, 151, 174
Werner, E., 127, 133
Wilson, D., 119n, 121
Wilson, M., 40, 62
Wooldridge, M.J, 125, 133, 135, 149

Y
Yeltsin, B., 66

Z
Zlotkin, G., 173

PHILOSOPHICAL STUDIES SERIES

1. Jay F. Rosenberg: *Linguistic Representation.* 1974 ISBN 90-277-0533-X
2. Wilfrid Sellars: *Essays in Philosophy and Its History.* 1974 ISBN 90-277-0526-7
3. Dickinson S. Miller: *Philosophical Analysis and Human Welfare.* Selected Essays and Chapters from Six Decades. Edited with an Introduction by Lloyd D. Easton. 1975 ISBN 90-277-0566-6
4. Keith Lehrer (ed.): *Analysis and Metaphysics.* Essays in Honor of R. M Chisholm. 1975 ISBN 90-277-0571-2
5. Carl Ginet: *Knowledge, Perception, and Memory.* 1975 ISBN 90-277-0574-7
6. Peter H. Hare and Edward H. Madden: *Causing, Perceiving and Believing.* An Examination of the Philosophy of C. J. Ducasse. 1975 ISBN 90-277-0563-1
7. Hector-Neri Castañeda: *Thinking and Doing.* The Philosophical Foundations of Institutions. 1975 ISBN 90-277-0610-7
8. John L. Pollock: *Subjunctive Reasoning.* 1976 ISBN 90-277-0701-4
9. Bruce Aune: *Reason and Action.* 1977 ISBN 90-277-0805-3
10. George Schlesinger: *Religion and Scientific Method.* 1977 ISBN 90-277-0815-0
11. Yirmiahu Yovel (ed.): *Philosophy of History and Action.* Papers presented at the First Jerusalem Philosophical Encounter (December 1974). 1978 ISBN 90-277-0890-8
12. Joseph C. Pitt (ed.): *The Philosophy of Wilfrid Sellars: Queries and Extensions.* 1978 ISBN 90-277-0903-3
13. Alvin I. Goldman and Jaegwon Kim (eds.): *Values and Morals.* Essays in Honor of William Frankena, Charles Stevenson, and Richard Brandt. 1978 ISBN 90-277-0914-9
14. Michael J. Loux: *Substance and Attribute.* A Study in Ontology. 1978 ISBN 90-277-0926-2
15. Ernest Sosa (ed.): *The Philosophy of Nicholas Rescher.* Discussion and Replies. 1979 ISBN 90-277-0962-9
16. Jeffrie G. Murphy: *Retribution, Justice, and Therapy.* Essays in the Philosophy of Law. 1979 ISBN 90-277-0998-X
17. George S. Pappas (ed.): *Justification and Knowledge.* New Studies in Epistemology. 1979 ISBN 90-277-1023-6
18. James W. Cornman: *Skepticism, Justification, and Explanation.* With a Bibliographic Essay by Walter N. Gregory. 1980 ISBN 90-277-1041-4
19. Peter van Inwagen (ed.): *Time and Cause.* Essays presented to Richard Taylor. 1980 ISBN 90-277-1048-1
20. Donald Nute: *Topics in Conditional Logic.* 1980 ISBN 90-277-1049-X
21. Risto Hilpinen (ed.): *Rationality in Science.* Studies in the Foundations of Science and Ethics. 1980 ISBN 90-277-1112-7
22. Georges Dicker: *Perceptual Knowledge.* An Analytical and Historical Study. 1980 ISBN 90-277-1130-5
23. Jay F. Rosenberg: *One World and Our Knowledge of It.* The Problematic of Realism in Post-Kantian Perspective. 1980 ISBN 90-277-1136-4
24. Keith Lehrer and Carl Wagner: *Rational Consensus in Science and Society.* A Philosophical and Mathematical Study. 1981 ISBN 90-277-1306-5

PHILOSOPHICAL STUDIES SERIES

25. David O'Connor: *The Metaphysics of G. E. Moore.* 1982 ISBN 90-277-1352-9
26. John D. Hodson: *The Ethics of Legal Coercion.* 1983 ISBN 90-277-1494-0
27. Robert J. Richman: *God, Free Will, and Morality.* Prolegomena to a Theory of Practical Reasoning. 1983 ISBN 90-277-1548-3
28. Terence Penelhum: *God and Skepticism.* A Study in Skepticism and Fideism. 1983 ISBN 90-277-1550-5
29. James Bogen and James E. McGuire (eds.): *How Things Are.* Studies in Predication and the History of Philosophy of Science. 1985 ISBN 90-277-1583-1
30. Clement Dore: *Theism.* 1984 ISBN 90-277-1683-8
31. Thomas L. Carson: *The Status of Morality.* 1984 ISBN 90-277-1619-9
32. Michael J. White: *Agency and Integrality.* Philosophical Themes in the Ancient Discussions of Determinism and Responsibility. 1985 ISBN 90-277-1968-3
33. Donald F. Gustafson: *Intention and Agency.* 1986 ISBN 90-277-2009-6
34. Paul K. Moser: *Empirical Justification.* 1985 ISBN 90-277-2041-X
35. Fred Feldman: *Doing the Best We Can.* An Essay in Informal Deontic Logic. 1986 ISBN 90-277-2164-5
36. G. W. Fitch: *Naming and Believing.* 1987 ISBN 90-277-2349-4
37. Terry Penner: *The Ascent from Nominalism.* Some Existence Arguments in Plato's Middle Dialogues. 1987 ISBN 90-277-2427-X
38. Robert G. Meyers: *The Likelihood of Knowledge.* 1988 ISBN 90-277-2671-X
39. David F. Austin (ed.): *Philosophical Analysis.* A Defense by Example. 1988 ISBN 90-277-2674-4
40. Stuart Silvers (ed.): *Rerepresentation.* Essays in the Philosophy of Mental Representation. 1988 ISBN 0-7923-0045-9
41. Michael P. Levine: *Hume and the Problem of Miracles.* A Solution. 1989 ISBN 0-7923-0043-2
42. Melvin Dalgarno and Eric Matthews (eds.): *The Philosophy of Thomas Reid.* 1989 ISBN 0-7923-0190-0
43. Kenneth R. Westphal: *Hegel's Epistemological Realism.* A Study of the Aim and Method of Hegel's *Phenomenology of Spirit.* 1989 ISBN 0-7923-0193-5
44. John W. Bender (ed.): *The Current State of the Coherence Theory.* Critical Essays on the Epistemic Theories of Keith Lehrer and Laurence BonJour, with Replies. 1989 ISBN 0-7923-0220-6
45. Roger D. Gallie: *Thomas Reid and 'The Way of Ideas'.* 1989 ISBN 0-7923-0390-3
46. J-C. Smith (ed.): *Historical Foundations of Cognitive Science.* 1990 ISBN 0-7923-0451-9
47. John Heil (ed.): *Cause, Mind, and Reality.* Essays Honoring C. B. Martin. 1989 ISBN 0-7923-0462-4
48. Michael D. Roth and Glenn Ross (eds.): *Doubting.* Contemporary Perspectives on Skepticism. 1990 ISBN 0-7923-0576-0
49. Rod Bertolet: *What is Said.* A Theory of Indirect Speech Reports. 1990 ISBN 0-7923-0792-5

PHILOSOPHICAL STUDIES SERIES

50. Bruce Russell (ed.): *Freedom, Rights and Pornography.* A Collection of Papers by Fred R. Berger. 1991 ISBN 0-7923-1034-9
51. Kevin Mulligan (ed.): *Language, Truth and Ontology.* 1992 ISBN 0-7923-1509-X
52. Jesús Ezquerro and Jesús M. Larrazabal (eds.): *Cognition, Semantics and Philosophy.* Proceedings of the First International Colloquium on Cognitive Science. 1992 ISBN 0-7923-1538-3
53. O.H. Green: *The Emotions.* A Philosophical Theory. 1992 ISBN 0-7923-1549-9
54. Jeffrie G. Murphy: *Retribution Reconsidered.* More Essays in the Philosophy of Law. 1992 ISBN 0-7923-1815-3
55. Phillip Montague: *In the Interests of Others.* An Essay in Moral Philosophy. 1992 ISBN 0-7923-1856-0
56. Jacques-Paul Dubucs (ed.): *Philosophy of Probability.* 1993 ISBN 0-7923-2385-8
57. Gary S. Rosenkrantz: *Haecceity.* An Ontological Essay. 1993 ISBN 0-7923-2438-2
58. Charles Landesman: *The Eye and the Mind.* Reflections on Perception and the Problem of Knowledge. 1994 ISBN 0-7923-2586-9
59. Paul Weingartner (ed.): *Scientific and Religious Belief.* 1994 ISBN 0-7923-2595-8
60. Michaelis Michael and John O'Leary-Hawthorne (eds.): *Philosophy in Mind.* The Place of Philosophy in the Study of Mind. 1994 ISBN 0-7923-3143-5
61. William H. Shaw: *Moore on Right and Wrong.* The Normative Ethics of G.E. Moore. 1995 ISBN 0-7923-3223-7
62. T.A. Blackson: *Inquiry, Forms, and Substances.* A Study in Plato's Metaphysics and Epistemology. 1995 ISBN 0-7923-3275-X
63. Debra Nails: *Agora, Academy, and the Conduct of Philosophy.* 1995 ISBN 0-7923-3543-0
64. Warren Shibles: *Emotion in Aesthetics.* 1995 ISBN 0-7923-3618-6
65. John Biro and Petr Kotatko (eds.): *Frege: Sense and Reference One Hundred Years Later.* 1995 ISBN 0-7923-3795-6
66. Mary Gore Forrester: *Persons, Animals, and Fetuses.* An Essay in Practical Ethics. 1996 ISBN 0-7923-3918-5
67. K. Lehrer, B.J. Lum, B.A. Slichta and N.D. Smith (eds.): *Knowledge, Teaching and Wisdom.* 1996 ISBN 0-7923-3980-0
68. Herbert Granger: *Aristotle's Idea of the Soul.* 1996 ISBN 0-7923-4033-7
69. Andy Clark, Jesús Ezquerro and Jesús M. Larrazabal (eds.): *Philosophy and Cognitive Science: Categories, Consciousness, and Reasoning.* Proceedings of the Second International Colloquium on Cogitive Science. 1996 ISBN 0-7923-4068-X
70. J. Mendola: *Human Thought.* 1997 ISBN 0-7923-4401-4
71. J. Wright: *Realism and Explanatory Priority.* 1997 ISBN 0-7923-4484-7
72. X. Arrazola, K. Korta and F.J. Pelletier (eds.): *Discourse, Interaction and Communication.* Proceedings of the Fourth International Colloquium on Cognitive Science. 1998 ISBN 0-7923-4952-0
73. E. Morscher, O. Neumaier and P. Simons (eds.): *Applied Ethics in a Troubled World.* 1998 ISBN 0-7923-4965-2

PHILOSOPHICAL STUDIES SERIES

74. R.O. Savage: *Real Alternatives, Leibniz's Metaphysics of Choice*. 1998 ISBN 0-7923-5057-X
75. Q. Gibson: *The Existence Principle*. 1998 ISBN 0-7923-5188-6
76. F. Orilia and W.J. Rapaport (eds.): *Thought, Language, and Ontology*. 1998 ISBN 0-7923-5197-5
77. J. Bransen and S.E. Cuypers (eds.): *Human Action, Deliberation and Causation*. 1998 ISBN 0-7923-5204-1
78. R.D. Gallie: *Thomas Reid: Ethics, Aesthetics and the Anatomy of the Self*. 1998 ISBN 0-7923-5241-6
79. K. Korta, E. Sosa and X. Arrazola (eds.): *Cognition, Agency and Rationality*. Proceedings of the Fifth International Colliquium on Cognitive Science. 1999 ISBN 0-7923-5973-9
80. M. Paul: *Success in Referential Communication*. 1999 ISBN 0-7923-5974-7

KLUWER ACADEMIC PUBLISHERS – DORDRECHT / BOSTON / LONDON

GPSR Compliance
The European Union's (EU) General Product Safety Regulation (GPSR) is a set of rules that requires consumer products to be safe and our obligations to ensure this.

If you have any concerns about our products, you can contact us on

ProductSafety@springernature.com

In case Publisher is established outside the EU, the EU authorized representative is:

Springer Nature Customer Service Center GmbH
Europaplatz 3
69115 Heidelberg, Germany

www.ingramcontent.com/pod-product-compliance
Lightning Source LLC
Chambersburg PA
CBHW060745150926
51501CB00003B/98
* 9 7 8 0 7 9 2 3 5 9 7 3 9 *